# AN ENGINEERING VIEW

# OF THE UNIVERSE

# VOL I

## ROBERT HEILMAN

## COPYRIGHT PAGE

## Table of Contents

# THE BEGINNINGS

Never thought I would write a book, especially about the state of Physics and the Universe, but, you never know. My name is Robert Heilman. When I graduated from High School I actually thought I was going to be an Accountant. So off to College I went and began learning Accounting. I was introduced to computers that were being integrated into the Business world. I learned Programming and computer code was like a breath of fresh air compared to Accounting. I loved the creativity and clear output results. To support myself while attending school, I landed a job at a local automotive parts manufacturer. Because Computers were starting to lead robotics into manufacturing, I found myself being pulled into Programming more and more. Getting my hands dirty, creativity, working with numbers, precision, and measureable results, I thought this was IT. But, as things go, the company was bought out and the staff reduced. I found myself without a job. So I began looking, But I was convinced that Accounting was out. The type of jobs I was looking for were far and few between, especially in my small Town. 60 miles away, in a city called Detroit, jobs were plentiful. Because of my computer and programming experience. I landed a job as an NC programmer and Design Studio support for an automotive supplier. I was exposed to many things I never knew existed! Computer Aided Design,

Computer Aided Simulations, Crash worthiness, weight reduction, Aerodynamics and wind tunnel testing, Strength of materials, Road Testing, In Process Testing, Analyzing Reports, Writing Reports, and many others; quality, design, production, inspection, cost reduction methodologies from the US, Japan, Germany and around the world. Things such as Statistical Process Control, Six Sigma, Etc. This was Engineering! It was a natural fit, especially with the integration of computers. Within a few years I was an Engineer and never looked back.

## ON TO PHYSICS

Why this book? Why now? Well, it would be easy to say I am of German ancestry and Germans have a long history in Physics. I MUST do this book, because cousin Einstein needs a hand, or some other nonsense. But the biggest driver is that I thought that using Engineering logic, I could help find Dark Matter/ Dark Energy. I started researching and reading Loads of articles, most by Physicists, some by anybody. I was just searching for the truth. As I researched, questions kept popping up that didn't seem right from an Engineer's point of view. So what does that mean? Engineers are physical science people: is it real? What are the mechanics: what does it do? How does it do it? What is the effect of what it does? How can it be made better? And then the actual Physical properties questions; what does it weigh? What are the dimensions? what is it made of? And finally. Test, Test, Test. Sometimes we run tests Just to make things Fail, so we know its capability and reliability. Most of the tests are video taped and we review and analyze. Some tests, especially Safety tests, are ran by independent Test Labs, with NO stake in the outcome and only care about running the test properly. So, people can tell much better Stories than I tell, and can Out-Calculate me all day long, but where the rubber meets the road in Physical Testing, I will own you. After a few thousand tests, you develop instincts for how things mechanically work and interact.  And finally, when I read a Theory, or about a test, or even just the movement and interaction of objects, this jumps out at me like it's in bold print.  This Book addresses the questions in the Universe as seen by an Engineer.

# CORRECTING HISTORY

But along the way, I ran into many unusual things; One of those things was well known Physicists (just a few really) claiming that certain things would not exist if not for Physics. Since Engineers design and build just about everything on the Planet and Space, my first thought was, what did we miss? Before I begin looking at the Universe, I would like set the record straight on a couple of items: One well know Physicist has videos of Just about everything and claims some things would not exist, if not for Physicists. I am an Engineer, I don't really need people recognizing my work or patting me on the back and telling me how big my brain is, but when someone else tries to take credit, that's going too far. (Ok, that and still wanting to find Dark Matter/Dark Energy). As you will see, the facts simply do not support these Physicist's statements: Physics has only been around since the 1600's if you include Galileo and Newton, neither of which graduated with a Physics Degree. Before someone whines that Physicist were actually in the Garden of Eden and helped God with Quantum Mechanics, consider this: The Telescope, the Microscope, the Vacuum pump, Electricity and the Light bulb, and much more, did not exist until after 1600, some well past. All of the Laws and Theories in Physics came after the 1600's. What would a Physicist even do before 1600? Play checkers? (come on, just kidding). Oh, and did I mention Degrees weren't handed out as well before 1600. The term *physicist* was coined by **William** WHEWELL (also the originator of the term "scientist") in his 1840 book, *The Philosophy of the Inductive Sciences,* but Engineering, on the other hand, has been around for arguably a few thousand years. Makes sense, since before the A.D.'s there were bridges to build, Great Cathedrals, buildings, roads and of course War machines and weapons. Physicists may have been in the Garden of Eden, but God was an Engineer! So, Isaac Newton did not have a Degree in Physics, but Marconi, Faraday and others did not even attend college. And the list goes on and on. Many of the known Physicists of the past were given Honorary Degrees only after some great accomplishment. Marconi won a Nobel Prize for Physics and then was given an Honorary Degree in Physics! Further, Let's look at the Steam Engine that was claimed by a Physicist would never exist if not for them, excerpt from the Encyclopedia Britannica: What very few *people know is that Heron was the first inventor of the Steam Engine, a steam powered device that was called aeolipile or the 'Heron engine'. The name comes from the Greek word 'Aiolos', who was the Greek God of the winds. But who was Hero? Hero of Alexandria, also known as Heron of Alexandria (10 AD – 70 AD) was a mathematician and engineer who was active in his native city of Alexandria, <u>Roman Egypt</u>. He is considered the greatest experimenter of antiquity and his work is representative* *of the Hellenistic method. An Engineer??The Greatest?? Hero described the construction of the aeolipile, which was a rocket-like reaction engine and the first-recorded steam engine; (although Vitruvius mentioned the aeolipile in De Architectura some 100 years earlier than Hero). The Vitruvius engine used air from a closed chamber heated by an altar fire to displace water from a sealed vessel; the water was collected and its weight, pulling on a rope, opened temple doors. Some historians have conflated the two inventions to assert that the aeolipile was capable of useful work. So, who was this Vitruvius? Maybe he was the missing Physicist who made the steam engine happen; Mainly known for his writings, Vitruvius was himself an Architect. In Roman times architecture was a broader subject than at present including the modern fields of architecture, construction management, construction engineering, mechanical engineering, military engineering and urban planning; Architectural Engineers consider him the first of their discipline, a specialization previously known as technical architecture. Frontinus mentions him in connection with the standard sizes of pipes. He is often credited as father of acoustics for describing the technique of echeas placement in theaters. - No Physics, but a whole lot of Engineering. Maybe we should go forward in time to where the this missing Physicists might hang out;*

**Thomas Savery** (/'seɪvəri/; c. 1650 – 1715) was an English inventor and engineer, born at Shilstone, a manor house near Modbury, Devon, England.  He invented the first commercially used steam powered device, a steam pump which is often referred to as an "engine". Savery's "engine" was a revolutionary method of pumping water, which solved the problem of mine drainage and made widespread public water supply practical. Received a Patent, 1698.  *Another Frickin' Engineer!  A Patent?? Wait, There's more;* **Thomas Newcomen** (February 1664 – 5 August  1729) was an English inventor who created the  first practical steam engine in 1712, the Newcomen atmospheric engine  He was an ironmonger by trade and a Baptist lay preacher by calling. He was born in England, to a merchant family and baptised at St. Saviour's Church on 24 February 1664. In those days flooding in coal and tin mines was a major  problem,and Newcomen was soon engaged in trying to improve ways to pump out the water from such mines.  His ironmonger's business specialized in designing, manufacturing and selling tools for the mining industry.

 **James Watt** was born in Greenock on 18 January 1736. ... Watt initially worked as a maker of Mathematical instruments, but soon became interested in steam engines. The first Working steam engine had been patented in 1698 and by the time of Watt's birth, Newcomen engines were pumping water from mines all over the country.

James Watt went on to become a large maker of  steam engines, but what is significant here is that by the time Watt was born in 1736, steam engines were already being used, 107 in use, *I looked it up.*  1700 years of history, all the way from ideas to

Steam Engines in use, including a Patent, and no Physicist in site (some Engineers though). So, is it a misinterpretation of the facts, or an honest mistake, that steam engines wouldn't be here without Physicists, or just a Damn Lie, as Mark Twain would say. Maybe by someone trying to sell Videos, or Books, or get Speaking Engagements, maybe even appear on TV. Well, you people decide, but I will tell you this, it took me all of 20 minutes to look up the facts. Anyone could have done this, not doing it is just sloppy work. ALSO The Laws of Thermodynamics were not published until 1850 and not completed until 1895, same thing with computers and Cell phones. Before either one, an Engineer patented the first circuit board. Try making a computer or Cell phone without a circuit Board! In fairness, most Physicists appear to be hard working, unassuming professionals. Just a few that are arrogant, misinformed, and act like they have big brains, and think videos, books, and speaking engagements make them scientists. But this book is not about Physics or Physicists, it is about the Universe. Is "observable proof" and Calculations acceptable proof for anything, or are those just a guide to narrow physical testing? Engineers think not so much on could be's or Theories, It either is or it isn't. We physically test over and over, we measure, we weigh and do hundreds of special tests, then review, evaluate, and re-test before we put our stamp of approval on it. And when Physicists couldn't figure out how to turn a wrench, we built the

CERN Collider for them, we build Space Stations for them, Space Ships(Shuttles) for them And the Hubble Telescope for them, in fact, almost all of the Telescopes for them. More Engineers have flown in Space than any other Profession, who better to comment on the Universe. Here is what the discovery of the Higgs Boson at CERN looked like: 13 Billion dollars and 10 years to get to this point, with no known uses for now. Now, besides building and maintaining CERN, Here is what some Engineers Accomplished in 1/3 the time and At a 10 Billionth

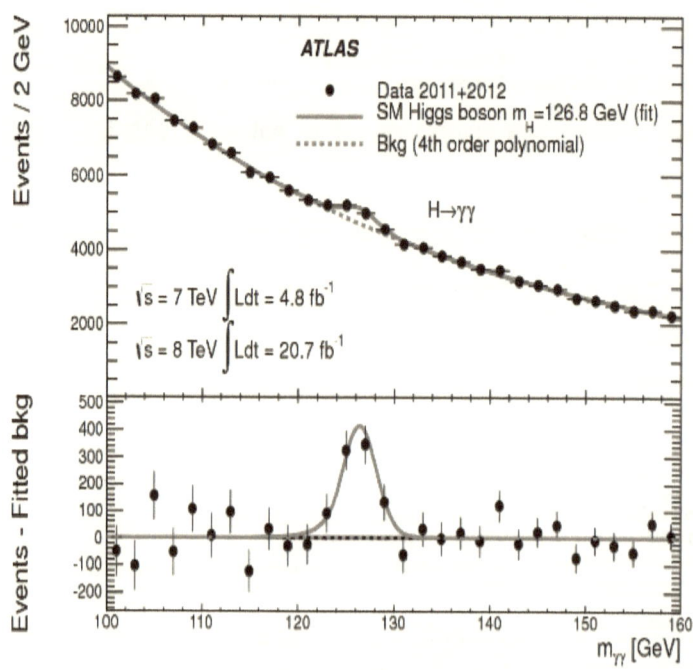

less cost:                                    Does
that mean Engineers are really responsible for every discovery in
the Universe.  No, but maybe, just maybe we helped a little.
And Please don't misunderstand, I believe Physicists are smart,
talented, and hard working.  I want you to succeed, BUT SHUT
UP AND GO TO WORK!!  No Videos, No Books, No public
appearances, No TV Shows.!!  This isn't a theatrical production,
this is Science.  Ok, done with my rant (sort of) So let's get on to
the Universe.

# THE GROUND WORK

But first, why would an Engineer even comment on the Universe. Because I have spent a lifetime learning how things move, go together and what makes sense and what does not. I guess you could say I have developed very good instincts for the Mechanics of things. I do not have to know or understand the entire Theory to know when a detail does not sound right. My attention to details, like most Engineers, has been required and refined. Matter in Space behaves much like matter on Earth. And just as important, logical thinking is the same too. The Scientific Method isn't Just for Earth. (Added a copy in the Glossary) The Devil (Truth) is in the detail and the steps to find that truth are the same. In fact, aside from the fact that it is harder to physically test in Space, the rest is the same. So why does it take so long, sometimes a hundred years, to make progress? Now I'll bet you think I am going to say "Physical Testing", that is correct, but not the only thing. Yes, it's time for a National Academy of Physics. For every Subject that I researched, there were books and videos galore. Anybody can say anything and it appears to be motivated by money, it doesn't matter if I know what I'm talking about, I can get a million hits! And the other thing that is opposite from Engineering is "Occam's Razor, Occam's Razor". Why is every solution complicated. Engineers actually get paid to reduce complexity in products, if we can make 1 part replace 3 that is a Gold Mine for us. So we tend to start with the simplest Solution first, then if necessary move to the next slightly more complex. Ok, who thinks String Theory is a very simple solution? You only need a PHD to understand it and what? 11 dimensions? So, this book is written really as questions, but in some cases I don't know what to say other than, "~~this is Stupid~~!" Oh, sorry, I mean I don't seem to get the logic. And on that note, let's look at some Einstein work to lay the ground work. We must have a remedial review of keeping and telling Time. We normal people learn to tell Time in grade school and we get it. But some Physicists, seem to think Time is negotiable. Time Dilation, Gravity Dilation, Twins Paradox, SpaceTime are all examples of Time being Manipulated. None of these subjects are true if you accept the Master Clock on Earth as the keeper of our Time. Sound pretty simple, right? In an Engineering Universe, nobody is touching the Clock.

# SPACETIME

## SO WRONG

Sometimes concepts are so simple that I wonder why People don't get them. Einstein treats Time as a fundamental Law of nature. Really? Is there A God who created Time? Does the Universe care about Time? No and No. Man created Time and we are the only one who cares about it. You could make a case that all Matter, Forces and Energies are the fundamentals and necessary for the Universe to be as is. But Time, No, the Universe could care less. Was there a conversation like, "Come on you Forces, we got to get this Big Bang done. We are on a tight schedule here!" It is Not a Force or an Energy or certainly not Mass. Time is a resultant, just as Man created it. A way of keeping a record of Events that happen. So simple. So a quick look at SpaceTime. Now let's look at the bending of Light. Look at this picture and you will see what is wrong:

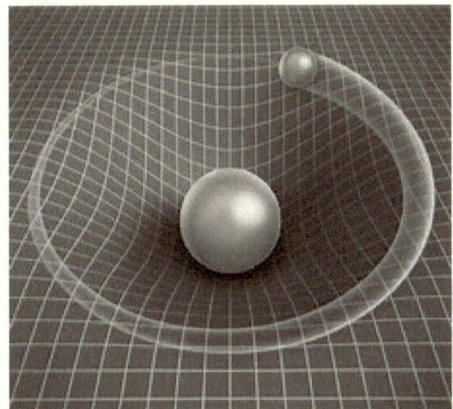
This shows Spacetime being distorted by A massive object with a Moon orbiting around it, this is not correct if SpaceTime is in fact 3D. I call this the bowling ball on a trampoline effect. Space would be distorted equally around a Sphere, but not like a bowling ball on a trampoline. There is no Gravity in Space to pull the ball down, so, I have put together a 3D picture of the real distortion zone.

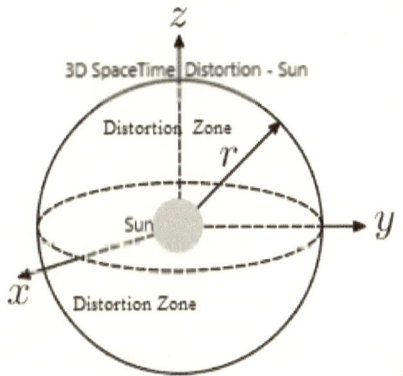

A Spherical Mass Object will distort Space

Spherically, not like the 2D pictures shown every where and above.

1. The outer circle represents the extent of the Sun's distortion, which is big enough to fit in our entire Solar System, remembering that all the Planets are held by the Sun. Before I go on, and after a lengthy discussion with my son (U of Michigan student), he did not think my 3D drawing was 3D enough. Since it takes time to model in 3D, I will walk you thru the concept. Lines of force (gravity) propagate from the Earth equally all around the sphere. Newton says that these lines go into Space and they pull on objects that come close enough to Earth.

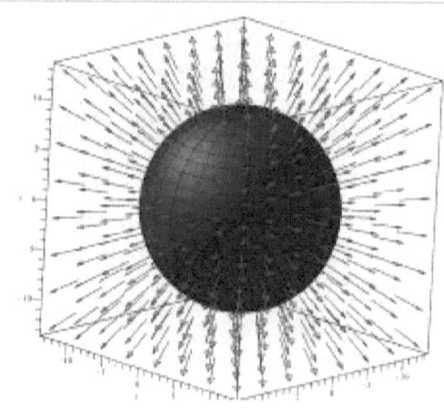

Einstein says no, no no. These lines actually pull (distort) on the matrix of SpaceTime like the pic below. Try to picture the pic above merging into this

pic 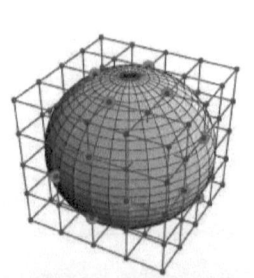 right.  on the Both effects extend well into Space.

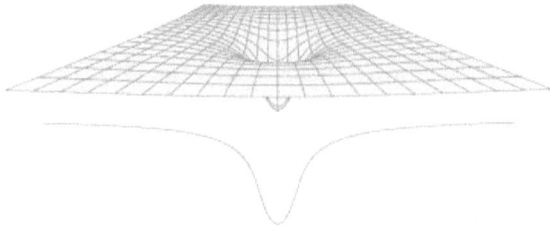

Instead of this 2D picture above, the true Distortion is Spherical around the Sun.  This makes perfect sense since we know Gravity is consistant around the Earth.  If you look at the true 3D distortion, questions arise.

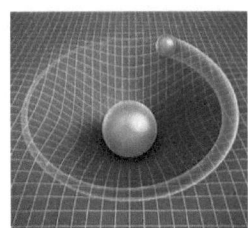

2. Einstein's distortion effect, while it can work, Causes a spherical distortion zone that favors only Circular orbits, elliptical orbits are a challenge

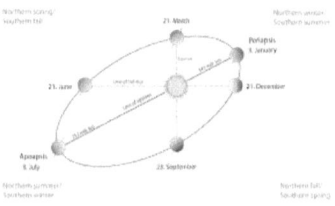

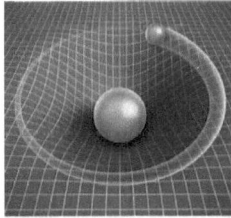

Earth's elliptical orbit doesn't work very well for Einstein's distortion concept. Newton's gravity works perfectly. Einstein is just using an Ether and calling it Space with Time thrown in.

3. See my 3D picture of the Sun to see the circular Distortion. Now let's see the Earth's orbit; Oh no, It's not circular at all. How can this be? The distortion of SpaceTime is equal all the way around a Sphere. Even if the plane of Rotation is angled, the offset is the same. Earth should be following the distortion. Newton's gravity could allow this, since Gravitational waves project from the Sun to the Planets. If the Planet picks up speed going towards the far point, this is the orbit we would expect to see, elliptical. Einstein's distortion of SpaceTime only allows for circular orbits, just like the picture. So Wrong. 4. And that's not all! The bending of light is questionable too. Look at a cut thru the center of the mass object.

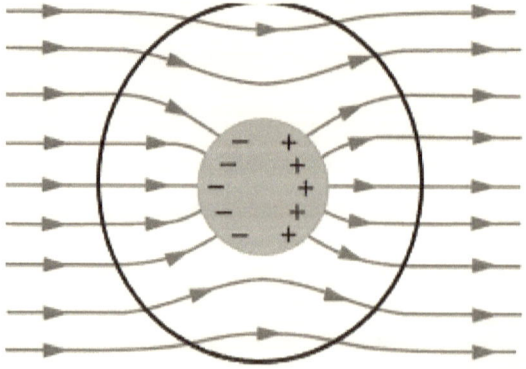

The black circle is the extent of the Distortion. Notice The middle line, there is no distortion at all. Seems Like a flaw in Einstein's Theory to me.

Newton's Gravity provides gravity all the way around. 5. I owe you a second Theory that can work with all the Spactime problems. This all sounds like an Ether to me. In fact, it appears Einstein just renamed the Ether to Space and added Time.

What is Einstein's Space made of that it can be Distorted. I would think that means it has mass. I believe this is true, but until it is Modelled, I can't say for sure. Also, this is what an Ether Would look like, and the propagation of Light is a quality of the Ether that Einstein killed. Just to end this SpaceTime Discussion, what you see here is 3D Space as it should be. The fabric of SpaceTime should be shown as 3D as well as everything in Space. Suddenly, things start looking a little differently. Are we to believe that not one Physicist can draw in 3D? If 1 picture is worth a 1000 words, then someone should learn to draw. Engineers Do. To wrap this up, here are some 2D pictures to Illustrate the point: Here is what the Ether was thought to look like during the Michelson-Morley

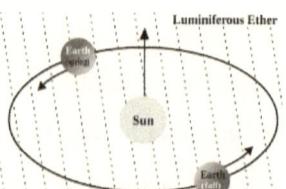

Experiment; ▼ ▼ ▼ ▼ ▼ ▼ ▼ ▼ ▼ ▼ ▼ ▼ ▼And here is Einstein's Distortion; Add the distortion to the Ether and these would be very similar. A coincidence? Maybe, but Einstein knew of the Michelson-Morley Experiment before Special Relativity with SpaceTime was published. SpaceTime is just the Ether with Time. Enough of the bending of light and Distortion from an Engineer's View There's more inconsistencies, but we shall forge ahead!

# TIME

I have rewritten this chapter 10 times, now 11, because the building blocks must be a solid foundation for things to come. Time is the #2 on my list, with the Ether being #1, but with Einstein doing parlor tricks with Time, #2 is gaining ground on #1. I really didn't want to write this Chapter, because it seems so simple. Elementary kids learn to tell time, but for some reason, Physicists think they can alter it. Sorry kids, the State requires us to teach Time Dilation and Gravity Dilation and Relative motion. And next week we will learn why your Twin Brother is younger than you! There are very few Facts in Physics, but this one should be undisputable. We are all Human, born under the same Clock, therefore any References to Time MUST agree with that Clock. End of Story. Time is a human invention to keep track of Events, eventually refined to years, months, weeks, days, hours, minutes, seconds. TIME IS A CONSTANT, maintained originally by the rotation of the Earth and now assisted from atomic clocks, ONLYI! No Time Dilation, No Gravitational Dilation. The Earth's Master Clock has NEVER seen Time Dilation just portable clocks. And since the Earth's Time Zones extend to the bottom of the Ocean and to the top the highest mountains, 12 am under the Ocean is the same as 12 am at Sea level and 12 am on top of the mountain. But, But, But say the Physicists, these things have been tested.

No, No, No Physicists! There has not been a proper test done yet. What about Heat or cold, think that could affect the test. How about atmospheric pressure or cabin pressure, and Radiation, electromagnetic waves, motion itself or Vibrations and humidity. Was the test in a controlled Environment and did items get torn down and inspected afterwards??? Throwing a few clocks in a plane is not a controlled test. Do you know that the atomic clocks keeping Earth's Time are in an Environmentally Controlled room to precision conditions, was that the case for the clocks in all the so-called tests. Were all the variables listed above, tested individually for their possible effects on clocks? No, I didn't think so. No real tests. It would be so much easier if we all on the same page doing the same thing. Since Time on Earth began WE Are the only known users of Time. Why would we change it?? Who are we trying to accommodate? Oh, I know, maybe somebody has a Theory they want To sell and the Theory just doesn't work right with this silly Earth Clock. Get used to it. TIME IS A CONSTANT. And, by the way, is more of a Constant than the Speed of Light, which only is Constant in a vacuum. Light can be slowed, diffused and stopped. So when an Eistein says that Time can slow down or you can live longer, maybe forever, You should be smart enough to say our Earth Clock is accurate to a Billion years, it does not slow down. Man-made clocks can slow down and be affected by variables. Everyone on Earth has been born against the Master and will die against the Master, nothing can change That. We are all Earthlings, wake up Physicists.! And since we fixed the Time reference, the logical Question would be, " Is there a SpaceTime?" NO, this is a word game. Is there Space, Yes. Is there Time, Yes. But they are not linked or intertwined. What Einstein does everytime he uses Time is to Change to a Reference Frame where it is possible. Against the Master Earth's Clock, most of his Theories Won't work, He just makes his own Time to prove his Theory. This is parlor tricks or card tricks or bait and Switch. No, there is no SpaceTime, or curved Space, Time Dilation, Twin Paradox, and much of anything Else. How ironic since Einstein trashed the Ether in favor of this sideways thinking. It would have been more accurate, If he just asked the Leprechauns or

the Unicorns

what Time it was. Now, we all got it, right? If a Physicist mentions Time other than a reference to our Master Clock, Your little BS detector should be sounding the Alarm! NOBODY TOUCHES THE CLOCK! A little more? Ok, but you owe me. I have seen so many explainations for Time my head is going to explode. "railroad tracks with loops and switches" "Blackholes are the end of Time, "infinite railroad tracks to the left and right with 0 at the center. Then a perpendicular set of tracks running thru the center." Are we talking about Time or Trains. Here is what Time Really, Really is; a Human invention. This is not the fourth dimension. IT can have multiple reference points along the measurement, A second, A minute, an hour, a day, a week, a month, a year, a lifetime, a Bigbang. All measurements along a Timeline. Man created Time to record events, nothing more, nothing less. Quit asking. Time would mean nothing to a God, it means nothing to the Universe. Only us!

# TIME DILATION

I really didn't know where to start the book, but this subject was a candidate, so many Flaws. Where to start? Since many people claim this has been tested several times, showing Time Dilation, let's Examine closer. If an Engineer did a test, and the Clock in motion showed a difference from a ground based clock, an Engineer would instantly think that something has affected the clock. Then all possibilities would be listed and reviewed, some possibilities could be eliminated. But for the rest, tests would be devised to simulate original As Tested conditions, even if it meant repeating the original test, but under more controlled conditions and better test equipment. Heat, Cold, Vibration, atmospheric pressure. Humidity, up/down, left/right movement, radiation, electromagnetic waves and magnetism. After all this, the clock would be specially prepared to filter out these possibilities (noise) and the tests ran again, paying close attention to in test operations, and any extra data available, like the Flight Recorder data for deviations to course, speed and altitude, all of which could affect time of flight. Because this was not done in any of the supposed tests proving Time Dilation, it would be equally valid to assume something else was affecting the clocks. This is what Engineers do, no stone unturned. But it gets worse; Let's take Gravitational Dilation. This is easy, meaning the higher you go, the faster clocks run. In the atmosphere, the less atmospheric pressure is Easy to Test, put the clock in a pressure chamber and increase and decrease the pressure and see how the clock is doing. Was this done? No! I am not in favor of playing with Clocks. Below the Ocean, at sea level, on top of a Mountain, 12 AM is 12 AM. And if it is not, FIX YOUR CLOCK. Time Dilation, FIX YOUR CLOCK. Imagine the complexity if you needed a computer to tell you the Time by how fast you were moving and by what your altitude is. But here is the real reason why Time Dilation DOES NOT EXIST. As you start moving your Clock starts slowing down at an ever increasing rate as speed increases. The problem is: Speed = Distance/Time. But, the faster you go the more that the Clock slows down(Time Dilation), the more the Clock slows down the higher your Speed increases because covering the same distance in less time = more Speed. No need to accelerate, It's an endless loop (Or a no solution Equation) More Time Dilation = More Speed=More

Time Dilation=More Speed, etc. etc. etc. And just to make this perfectly clear; if a Spaceship flies a mile in a minute, it appears to be going 60 mph, But Time Dilation said the clock slowed down. Instead of a minute to go a mile, it only took 55 seconds, so the speed was really 65 mph, but if the speed was higher, Dilation would cause the clock to have slowed even more, which would then cause the speed to go up, an endless loop. By this logic, I guess Infinity Speed is possible. And guess who didn't see this little loop problem? Or should it be called the Time Dilation Paradox? Wouldn't One Universal Clock be so much easier. Here that? National Academy of Physics. Ifyou allow people to create their own Universe, it makes it increasing hard to find the real Truth. I am not a big supporter of the Nobel Prize Committee, but they never awarded the Nobel Prize to either General Relativity, Or Special Relativity, Nailed it! Simplify, Simplify.

Examples of 2D Space in books, videos, TV, Internet. This not what 3D Space looks like.

# THE AETHER

## An Intro

I am not a Physicist, I am an Engineer. I don't speak the language. Complex math? Just reading it gives me a headache. But what I CAN do, is figure things out; How they move, why they move, when they more. And probably more importantly, I can see how they are built, how they fit together, what makes sense, what does not, and seeing all the possibilities. Physically test it, over and over, and Document everything. I see the big picture, there must be a game plan. Working bottom up is not efficient(Quantum/Particle Physics); If you buy 10,000 Lego's, yes you can make anything, but if you only needed 100 then you just wasted a lot of time and money, thereby slowing down the whole process, for something you don't need or want. Physics, today works on everything with no agreement to how the big picture works. Physics is trying to explain Gravity, Dark Matter/Energy, Inflation, Propagation of electromagnetic Waves, Curvature of SpaceTime, Relativity itself, and God knows why they can't tell Time. And there is still support for an Aether (Ether). The Theory seems to be; if we discover enough sub-atomic particles, we can explain anything! This is why you are arguing about things 100 years after they were published. Have a

| | nanoseconds gained, predicted | | | nanoseconds gained, measured | difference |
|---|---|---|---|---|---|
| | gravitational (general relativity) | kinematic (special relativity) | total | | |
| eastward | +144 ±14 | −184 ±18 | −40 ±23 | −59 ±10 | 0.76 σ |
| westward | +179 ±18 | +96 ±10 | +275 ±21 | +273 ±7 | 0.09 σ |

top plan and work to support it with clearly defined steps and look to communize, unify and borrow always. See the Big Picture. Name one thing in Engineering that is still being debated from 100 years ago. Physical tests solves these problems. Learn to love Data, it is he real Truth, no Theories needed. Again, how many Theories do you see in Engineering. No one or, a God, gave us Test Equipment, we made it ourselves! Think of it as a quest for Truth. I can't do Differential Equations, but I will spreadsheet you to death! (funny, Physics called it Statistical Physics) I am a data Monster. Finally, I can make you money doing what I do, oh wait, I guess that's irrelevant here. Looks good on the Performance Review though. But, it does show a level of efficiency; cheapest, lightest, strongest, NO WASTED MOTION. Almost like a Unified Theory. So, I bring a different perspective to the Universe. All I really wanted to do was do a spreadsheet, find the Dark Matter/Dark Energy, call it a day, move on. What I found was that every subject that I researched had unresolved questions (flaws?). Gravity, Speed of Light, the Ether, Black Holes, Expansion, Etc. And since everybody gets a say, finding the best or right data is very difficult. Picking the most important place to start with was a coin flip. All I cared about was Dark Matter/Dark Energy. The Ether solves both so let's go. The name AETHER is the Old traditional name for something in Space responsible for many uses. Later updated to the Ether. I don't care what name we use, it is the concept that shows promise. It may be Dark Matter/Energy. It may be Inflation. It may be Gravity. It may be the limiting factor to the speed of light. Or all of the above! I try to be honest, so yes, this is a WAG, but a missing piece to the Universe. This would simplify things enormously.

# THE AETHER
## 3D Space

So, from an Engineering perspective here is my #1 issue and the
#1 issue in the Universe – The Aether, is it or isn't it. Why is it
#1? Because this could answer so many questions; Speed of
Light, Gravity, Inflation, Dark matter/Energy. This is Big
Picture. I explain each issue that it can solve in more detail later,
but quickly, An Ether explains the Speed of Light Speed, as this
is the maximum speed the Ether can handle, Gravity? An Ether
running thru the Universe would contact all the matter in the
Universe and may be carrying the Gravity affect. Inflation? An
Ether carrying the Gravity affect would be Negative, and if it
was more Negative towards the outside of the Universe, Matter
would be attracted to the outside, and if it was significantly more
negative towards the outside, matter would speed up as it travels
towards the outside. Dark Matter? The Ether needs to be made
of something and that something may be just some regular
matter that is thinly distributed thru the entire Universe, a Dark
Matter effect. Remember that everything from a molecule on
down can't be seen. Dark Energy? Again, if something is
spread across the Entire Universe and carrying a Negative
charge, adding up the Energy in this charge may equal Dark
Energy. This is potentially a Huge Discovery. Why Physics
aren't hot on this issue just baffles me. The Ether has the

potential to help unify the Universe. A little History: Physicists were hot on the trail of The Ether in the late 1800's, when things took a turn for the worse, as Physicists decided to run an experiment around 1900 to look for an Ether, The Michelson-Morley Experiment. From an Engineering perspective, THIS IS A FLAWED TEST. I don't know much about Physics, but I do know about Testing and Test procedures. One of Physics most famous Experiments, this illustrates perfectly the difference between Engineers and Physicists. The Physicists ran ONE experiment, the first time it had ever been ran, and concluded That there was no Ether and then tried to move on to some other concept. Engineers would have ran this test into the ground, over and over, Changing the variables, Changing the test setup, finding how good is good or how bad is bad. What could we do to make it pass, what could we do to make it fail. Challenging the assumptions, utilizing different materials (test was ran with Light, why not run it with Radio waves?) Engineers are relentless. If this experiment was so critical, it probably would have been moved to another facility and re-ran again. In the end, when we say it is so, we have reams of data to support it. 100 years later and we are still arguing about just about everything in Physics. Think about it, when was the last time anything in Engineering was questioned? Not a bad record considering we design and build everything on the Planet! So, the Ether was thrown out on one first time experiment: The concept of an

Ether has actually been around for hundreds, almost thousands, of years. The ancient Greeks, before the first Century, wrote about some mysterious Aether above the Earth. This continued thru the centuries into Medieval Europe where they believed the essence of the Ether had healing properties. Up to Isaac Newton. Here is Newton writing about the Ether: Aether and gravitation (Borrowed from Wikipedia).

Sir Isaac Newton. Aether has been used in various gravitational theories as a medium to help explain gravitation and what causes it. It was used in one of Sir Isaac Newton's first published theories of gravitation, Philosophiæ Naturalis Principia Mathematica (the Principia). He based the whole description of planetary motions on a theoretical law of dynamic interactions. He renounced standing attempts at accounting for this particular form of interaction between distant bodies by introducing a mechanism of propagation through an intervening medium. He calls this intervening medium aether. In his aether model, Newton describes aether as a medium that "flows" continually downward toward the Earth's surface and is partially absorbed and partially diffused. This "circulation" of aether is what he associated the force of gravity with to help explain the action of gravity in a non-mechanical fashion. This theory described different aether densities, creating an aether density gradient. His theory also explains that aether was dense within objects and rare without them. And particles of denser aether interacted with the rare aether they were attracted back to the dense aether much like cooling vapors of water are attracted back to each other to form water. In the Principia he attempts to explain the elasticity and movement of aether by relating aether to his static model of fluids. This elastic interaction is what caused the pull of gravity to take place, according to this early theory, and allowed an explanation for action at a distance instead of action through direct contact. Newton also explained this changing rarity and density of aether in his letter to Robert Boyle in 1679. He illustrated aether and its field around objects in this letter as well and used this as a way to inform Robert Boyle about his theory. Although Newton eventually changed his theory of gravitation to one involving force and the laws of motion, his starting point for the modern understanding and explanation of gravity came from his original Aether model on gravitation. Aether and light (Borrowed from Wikipedia): Main article: *Luminiferous aether. The motion of light was a long-standing investigation in physics for hundreds of years before the 20th century. The use of aether to describe this motion was popular during the 17th and 18th centuries, including a theory proposed by Johann II Bernoulli, who was recognized in 1736 with the prize of the French Academy. In his theory, all space is permeated by aether containing "excessively small whirlpools".*

*These whirlpools allow for aether to have a certain elasticity, transmitting vibrations from the corpuscular packets of light as they travel through. This theory of luminiferous aether would influence the wave theory of light proposed by Christiaan Huygens, in which light traveled in the form of longitudinal waves via an "omnipresent, perfectly elastic medium having zero density, called Aether. At the time, it was thought that in order for light to travel through a vacuum, there must have been a medium filling the void through which it could propagate, as sound through air or ripples in a pool. Later, when it was proved that the nature of light wave is transverse instead of longitudinal, Huygens' theory was replaced by subsequent theories proposed by Maxwell, Einstein and de Broglie, which rejected the existence and necessity of aether to explain the various optical phenomena. These theories were supported by the results of the Michelson–Morley experiment in which evidence for the motion of aether was conclusively absent. The results of the experiment influenced many physicists of the time and contributed to the eventual development of Einstein's Theory of Special Relativity.*

**The Ether – Michelson-Morley The Big Flaw** This is my biggest disappointment with past Physics. It is almost inconceivable that people with PHD's did not know the difference between an Experiment and a Test. Even when it was labelled as an Experiment. Just so everyone knows, here it is: A test isn't an experiment, and an experiment isn't a test. Experimentation is done without a fear of failure or expectation of outcome. Testing is done with an expectation of a winner. Experimentation leads to new things, while testing validates assumptions. There is a difference. I will now point out the Flaws of this Experiment if it is considered as a Test. Yes, Flaw #1, it was not a test, but an Experiment and labelled as such. I have no problem with doing experiments. It's like a fishing expedition trying to gather data to refine a Theory. This is part of the preliminaries. A test is The Big Show. You are trying to validate something, mostly A Theory, but it can be a piece of the puzzle too. Engineers plan, detail and obsess to pass a Test. Do we fail? Yes, occasionally. Do we make mistakes? Sometimes. But major Tests cost Money. God help you if you didn't do your homework. One thing Engineers don't do, is call an Experiment a Test. Throw out the whole Ether Theory on 1 experiment, ran 1 time? That's Crazy. Then teach the results as a Test for a Century afterwards, UNBELIVABLE. NOTE: After studying the Michelson-Morley Experiment, It appears the experiment was an attempt to prove there was an Ether. I say appears, because as with everything in Physics, there are different versions of every story. The premise was: The Earth traveling through an Aether would create a wind, an Ether wind,

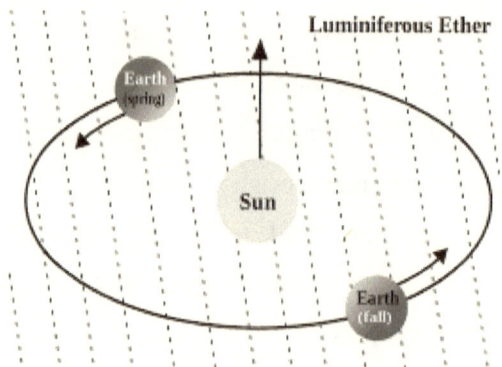

that would cause a light beam traveling in the opposite direction to be affected or slowed slightly by this wind, and with the wind, to go faster, and if perpendicular not affected, and, of course, this effect could be measured. And if true, it would mean an Ether existed. But if not, then maybe no ether. FLAW #1 Since when do the people conducting the experiment get to makeup properties for the thing they are testing. The Ether may not even be affected by the Earth passing thru it. You don't assume anything, YOU TEST. NOBODY EVER TESTED FOR A WIND, IT MIGHT NOT EXIST! So the Premise is Flawed and unsubstantiated or more correctly, this Ether wind was never tested for independently. FLAW #2 EVEN WITHOUT AN ETHER WIND, THERE COULD STILL BE AN ETHER, this possibility was never reviewed and addressed. Never talked about, never written about. In fact there was a third option that has never been addressed: The Ether itself is the limiting factor to the propagation of electromechanical waves. The results COULD be explained as, Wind or No Wind, an Ether propagating at the speed of Light, the same as was assumed for empty Space. The test results would be the same with an Ether or no Ether. What is the point? To do this test today would be difficult even by today's standards and precision measuring devices. Trying to measure a difference in the speed of beams of light to a millionth of a second in the late 1800's would have been impossible. As evidence of how improbable the accuracy of these test results would have been, if you examine the test setup closely, you will see that it is on a large tabletop. The beams of light had no more than a few feet to go until they hit the detector. A few feet to detect the difference in the speed of 2 beams of light going 186,000 miles a second. Like I stated, a millionth of a second. Now, before the Physicists start kicking and screaming about the fact that the beams were not timed, but the wave patterns from each beam were compared and it would be easy to see if the waves were in "phase" or "out of phase". Do you really think you can see a millionth of a second "out of phase" condition? I already talked about the reliability of observable proof", as compared to measuring. This brings us back to my point that in the late 1800's, the equipment would not have been sensitive enough or sophisticated enough to get irrefutable results. But Einstein was trying to sell his Theory of Special Relativity and the Ether was in his way to selling SpaceTime. Einstein jumps on the test results, even though it was ran only once with a flawed Premise and results. Why can I see the inconsistencies and Einstein didn't or did he? So Einstein thinks "close enough" and throws out The Ether Theory. Most Physicists leaned towards "does not exist" with Einstein dismissing the Ether in Special Relativity. Now to be fair to Einstein, he had nothing to do with the design or execution of the Experiment or the accuracy of the results. And Einstein was one of the Physicists who believed there was a particle in light and he had no clue that years later the mass would get thrown out and thereby change the assumption, that the Ether wind would affect light. How does a wind affect a massless particle? FLAW #3 ( I truly believe that If Michelson-Morley knew there was no mass in light, they never would have ran this experiment.) So, Einstein wasn't a big opportunist, just merely lucky this experiment took place when it did. The whole Experiment just assumed the Ether to be there or not there. But they missed the third and most important Possibility: It was there, but unaffected by the wind, OR the wind did not even exist. Its makeup can only support electromechanical waves at the speed of light, and further, it actually may be the limiting factor in the speed of light! They overlooked this possibility, Physicists overlooked this possibility, Einstein overlooked the possibility. Since we know Light has a massless particle (What??) It could not be affected by an Ether wind. (FLAW #3), it was assumed it would be affected. The test results would be the same, Ether or not. The experiment proved NOTHING. In 1905, influenced by this experiment, Einstein lays waste to the Ether in Special Relativity. Then, in 1915 publishes General Relativity, AND, wait for it, here is his comment; *According to the general theory of relativity space without ether is unthinkable; for in such space there not only would be no propagation of light, but also no possibility of existence for standards of space and time. But this ether may not be thought of as endowed with the quality characteristic of matter, as consisting of parts ('particles') which may be tracked through time.*

(*Albert Einstein, 1928, Leiden Lecture*) Anything to prove Special Relativity, trashing the Ether, even though Einstein clearly thought there was one. Here we sit, No Ether or even research into it. One could conclude SpaceTime would not exist if the Michelson-Morley Experiment was done correctly.

Michelson-Morley Experiment Setup.

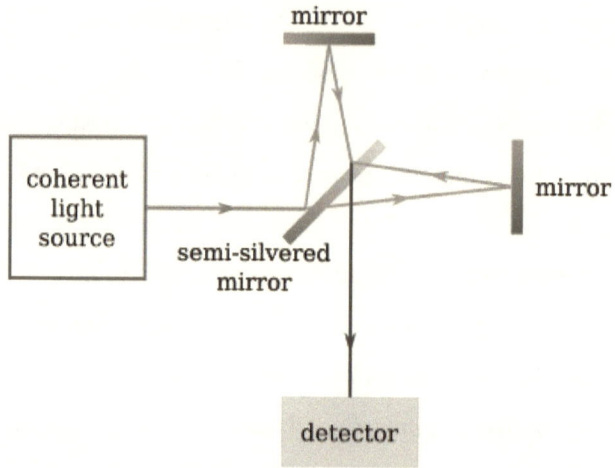

**Michelson-Morley Experiment.** This is so important, I feel that I have to make it as clear as a bell; The Michelson-Morley Experiment was a flawed Experiment for 3 reasons,

1) The Premise of an Ether wind being generated by the Earth passing thru the Ether is UNPROVEN, A WAG.

2) Light has no mass, therefore, could NOT BE DISTURBED by a wind.

3) The equipment of the day or that they had was not Precise enough to get any verifiable evidence of ANYTHING. The Difference in speed of 2 beams travelling 3-4 FEET at 186,000 MILES a sec., a difference of maybe a millionth of a second. AND FOR THESE 3 REASONS, THE EXPERIMENT RESULTS SHOULD NOT BE CONSIDERED!! But, back to an Engineer's view; A 3D matrix in some geometric shape (cubical?) made of particles, like electrons and/or protons or

even sub-atomic particles is extending thru-out the whole Universe. The logic is very solid, The energy of the Universe (electromagnetic waves) needs something to move along. In the Big Bang, 73% or more of the matter produced was Hydrogen. So it makes perfect sense that Hydrogen with free electrons, utilizing a type of covalent bonding technique, could form a matrix or Ether thru-out the expanding Universe. That is my WAG. Quantities of Hydrogen are there, types of bonding techniques are there, time needed to develop is there, sources of free electrons are there. The building blocks are there. For Quantum particles, this would be a perfect application. 1) **Newton** thought there was an Ether.

2) **James Maxwell**, father of electromagnetism,     believed there was an Ether

3) **Tesla** believed an Ether was there.

4) Of course **Lorentz and Poincare** believed it was    there. Einstein, knew Maxwell's work and knew Lorentz, and came up with a similar concept, but called it SpaceTime, with a "fabric" of space, but never defined what made the fabric. 5) Even Einstein came around. The list goes on. There is something out there. Moving on. . .     Around 1900, Physicist, Hendrik Lorentz with the help of Henri Poincare proposed a Theory of the Ether commonly known as the Lorentz Ether Theory (LET). This was the leading candidate for explaining Space at the time, but then along came Einstein with Special Relativity (SR).

Einstein believed that the Ether was not needed, as SR worked just fine without it. But of course, he introduced SpaceTime that had a fabric, but no definition of the fabric. The popularity of an Ether Theory declined and was never accepted. Although in doing research, I found Ether Theories are still alive. I looked at eight such Theories, most by Physicists. So to help, here is a list of what each leading Theory (LET)/(SR) will provide:

| <u>LET</u> | <u>SR</u> |
|---|---|
| *-A medium for propagation of Electromechanical waves | *-Clocks slowdown in Space. |
| *-A reason for Light Speed (max speed Ether will allow) | *-Reference frames *-SpaceTime |
| *-Support for a pushing Gravity explanation for Inflation. | (Fabric in Space). *-An *-Twin Paradox. |
| *-An explanation for all Dark Matter | *-Time Dilation |
| *-Partial if not all explanation of Dark Energy. | |
| *-Allows possibility of mass objects faster than light | |

Now, if you had to figure out the Universe, which Theory would you choose? Clearly LET, in fact, throw out the manipulation of Time in Einstein's SR and there is almost nothing. And, in a real Irony, Einstein actually proves the Ether concept by introducing a Fabric of SpaceTime. How do you have a fabric with no fabric? Einstein never clearly defined what made up the fabric and whether it was 2D or 3D, but a 2D fabric in a 3D Universe would be just ridiculous. And Einstein stated that space was 3D. As for the rest of SR; Sorry, but time cannot be manipulated, IT IS A CONSTANT, it is kept by the Earth's rotation (atomic clock) and hasn't missed a beat in, oh say a billion years. (See TIME chapter) SR has so many flaws and just plain unusual concepts, but more later. Predictably, Physicists chose Einstein's SR as their path. ?? But Even if LET is flawed or there is a better Ether Theory out there, it helps to unify the Universe.

Einstein/Lorentz I would choose to pursue LET first. So, sorry, for all the background info on the Ether, because it really doesn't matter who did what, or who said what. The real question is, Can an Ether exist and what makes me think it can. It is simple really, Occam's razor. In researching, I found that Hydrogen is the most abundant element in the Universe. In addition, Stars in burning, produce ions of hydrogen and strip electrons. And if that isn't enough, high energy Waves, gamma and x-ray, in collisions with particles, can strip off electrons. In fact, this happens in the Earth's Ionosphere every day. But not just Earth, everywhere in the Universe. Millions of Billions of free electrons to go with the most abundant Element in the Universe, Hydrogen. It is a short walk to see these two combining in free Space to form a matrix. The Hydrogen supplies the structure, electrons supply the negative (Gravity like) charge. I know what you are thinking, when atoms bond there is usually no charge, protons and electrons equal out. But in free Space, especially if free electrons are abundant, a positive charged proton could form an attraction for more electrons because, let's face it, in free Space it is the only game in town. Even if the proton shares a charge of +1 with eight electrons -8, a bond of +.1 to -1 is the only bond available. This would also explain why the Ether has always been thought of as fluid and stretchable, and flows around mass objects to form Gravity. This adds mass to the Universe (Dark Matter) Energy (Dark Energy) and can support

Inflation. Is the matrix cubical? or spherical? or something else? TBD, but I picked cubical because it makes it easier to share electrons at the corners. All the components are there!

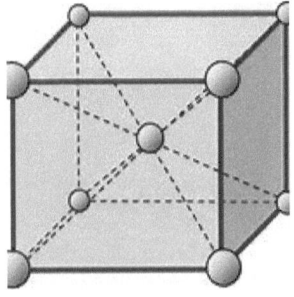 Even though using Hydrogen and free electrons makes sense I am still looking at different ways of bonding and availability of elements, ions or particles to make this work. Sub-atomic Particles could do this, but right now that is a little outside of my Job Classification. One thing that does look interesting is Isopropyl Cyanide, creating long string structures in Space. Carbon would be excellent to build an Ether with considering it is what used in Nano structures. But Carbon is considerably more massive than hydrogen so the mass of the Universe must be looked at. It is Hydrogen for now. And more on the specifics of Nature building the Ether in a later Chapter.

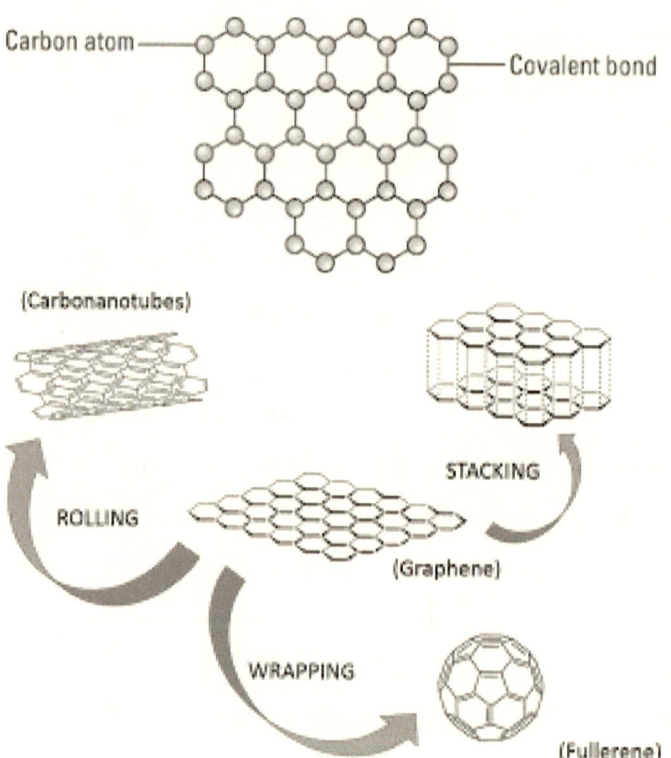

Carbon atom —

— Covalent bond

(Carbonanotubes)

ROLLING

STACKING

(Graphene)

WRAPPING

(Fullerene)

# LIGHT CLOCKS

I never thought I would be writing a chapter on Clocks, but so many people have a misperception of Clocks, I thought I may as well start at the beginning. Here is a definition: The light clock is a simple way of showing a basic feature of Special relativity. A clock is designed to work by bouncing a flash of light off a distant mirror and using its return to trigger another flash of light, meanwhile counting how many flashes have occurred along the way and displaying that on the clock. Pretty simple and it is, until the Spaceship and the clock in it begins accerating to the Speed of Light. The light moves up and down, but as the Ship moves laterally at the speed of light, the clock moves laterally also at the Speed of Light. The result is that the light looks like it is moving at an Angle, 1,A,3 This is illustrated by the Drawing below:

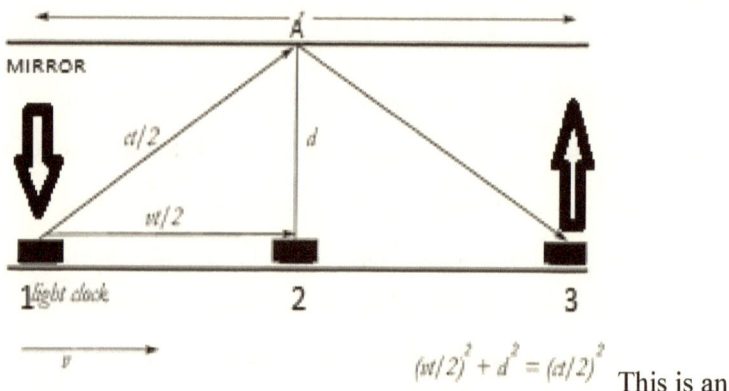

$$(vt/2)^2 + d^2 = (ct/2)^2$$ This is an imaginary clock that is used to illustrate Einstein's Time Dilation. Einstein had nothing to do with developing this drawing, but in practicality it does show how Time Dilation works, but sadly it actually shows why there is no Time Dilation. I am an Engineer, I can't spot a flaw in a complicated Equation, or word-games and abbreviations leave me wondering, but when it comes to how things work or move, you better not make a mistake. Such is the case here. Here's how this works and why Time Dilation doesn't. Block #2 represents the rest position of the Clock. Rest position means The Ship is not moving. The block 2 emits light and it goes to point A, a mirror, and is reflected back to a detector in block #2, this causes the clock to click 1 increment. Real simple; Light up, light down, click the clock. Now the Ship Starts moving all the way to light speed and the Einstein followers want us to believe this; now use Block #3 as the start, Light will be emitted at #3 And at the same time the Ship, flying at light speed, Will fly to the right shifting the

clock over. #2 will now move to the #3 position and point A is now above Old #3 and light reflects off the mirror and starts back Down. Again the Ship flying near light speed, moves #1 under the light and the light hits the detector at #1. Now the smart guys want us to believe that the light is following the diagonal paths of #3 to point A to #1. This is further than just going straight up and down. Since it takes longer to cover the diagonal distance, the clock runs slower, Time Dilation. Let an Engineer Explain what really happens. I told you the story as It really happens hoping you got it, but in case you Didn't here it is: The emitter is set to the perpendicular Position to emit light. It does not matter if is going 1 mph or 186000mps the light is propagated vertically. The mirrors are set to reflect vertically. So, light is only going where it is pointed, straight up, straight Down. Sure the moving ship is moving new mirrors above it and new detectors below it but it doesn't Care. Straight up, straight down. Just like the rest Position. Sorry, Einstein can't fool an Engineer. NO TIME DILATION!! NO CONTRACTION.

# THE SPEED OF LIGHT

I Was going to save this for the next book, but too many people have requested me to comment. Oh God, here it goes: Around 1900, Lorentz and Poincare were working on Relativity and the Ether or Aether. This Ether was everywhere in the Universe and was supposed to support the propagation of electromagnetic waves. Well, along comes Einstein with Special Relativity and a bunch of Physicists who can't tell real time, and trashes the Aether concept, as it is not needed in Special Relativity per Einstein, but then comes out with his own version of the Ether and calls it SpaceTime. But then conveniently doesn't define what this fabric is made of. A fabric running all through space? But Space is 3D and a 3D fabric sounds a lot like an Ether to me. But Einstein overlooked a very important point; the Ether may limit the speed that light can propagate! Nobody talks about this possibility. As evidence of this, waves need a medium to propagate in: Sound waves need the air, Ocean waves need the water, shock waves need water, air or land. It is logical to assume electromechanical waves need something too. Now I am not the first to say this, I don't like to take credit for things I didn't do, but is it pure coincidence that all electromechanical waves propagate at the speed of light? Low Energy radio waves to high energy X-rays and Gamma waves, quite a coincidence. I make a living out of paying attention to how things move and this doesn't seem right. So, the real speed of light limit may be that the Ether can only propagate light that fast. This fits nicely with speed at low levels being additive, but not near the speed light. Shoot a bullet from a flying jet and the speed of the jet becomes added to the speed of the bullet. Shoot a beam of light from a jet and it still only goes the speed of light. The Ether is what is causing this affect. WAG. BUT, DO NOT get light confused with mass objects. Only electromechanical waves have this limit. Ever sit by a Lake or an Ocean and see waves reaching the shore, then a speed boat goes racing by, this is space. Waves propagate at one speed and starships are capable of much higher speeds. Suppose you have a 1 mile long race track. And you make 1 foot high walls around it. You then fill this with 1 foot of water. You then drive down the track and time it. You drain it, and fill it with sand. Drive it and time it. Again drain it. This time don't fill it with anything, just air. Drive it and time it. It becomes apparent that resistance to movement makes a difference. Do the same tests again, only this time put a much bigger engine in the car. Again, it becomes apparent that more power means more speed. I know I have lost the physicists by now, but I will continue and hope the light goes on. So now we take the car(space craft) into space. What do you know, there is no resistance to movement. The car will go as fast as the engine can move it. Yes, we may have to overcome tiny inertia, but with no resistance to movement, that will take very little. NOTE: I know there is some resistance and acceleration, but way less than Earth. This is like saying a massless particle, an oxymoron. Eventually a combustion engine can only go so fast, so we switch to a rocket engine. Again no resistance, so we can accelerate to hundreds of thousands of miles an hour. Since there is no resistance to movement we could coast for years, no power required to maintain current speed. So, what the heck, let's go for the speed of light. Since no resistance to movement, all we need is an engine that can provide speed of light thrust, similar to electron beams or laser beams or Tachyons, they do exist, right? No inertia to overcome with zero resistance. All the mumbo jumbo about massive objects nearing the speed of light becoming infinitely massive, requiring infinite energy to move it, is a load of fertilizer. Who started this thing about mass and energy Zero resistance means very little energy to move it, in effect, becoming a zero mass object. The speed of light is just a number, not a stop sign. Just as Einstein's $E=mc^2$ falls apart with zero mass, so does Newton's $F=MA$ does too. So what this means per Einstein is that any energy faster than the speed of light would be enough to move a massless spacecraft to faster than the speed of light. Newton's Law in a strange way agrees. Acceleration is the force required to overcome a resistance to movement. Take away the acceleration needed in Space, and the formula becomes $F=m \times speed$. Take away the mass and $F=speed$. The force needed to go faster than the speed of light is just a thrust engine that can produce faster than light speed thrust. The real issue is not whether an object can go faster than the speed of light, it is all about having a thrust engine that can provide faster than light thrust. So simple my Grandmother could figure this out. In all fairness, by proposing an Ether matrix in Space, there would be some resistance to movement on the spacecraft in the form of gravity generated by the spaceship distorting space. But, we can go faster than the speed of light, but I'll save that for the next book. Also, I am really not talking about anything new in this Chapter. Theories of electromagnetic waves (light) needing a medium for propagation have been around for a hundred years. Ask Tesla, Oh wait. I am merely highlighting the importence that this discovery would be. This could change Physics. Even though I believe the Ether exists, Physicists are right in waiting for absolute proof. Sounds funny, but they are right. So sell The CERN collider for scrap and let's start testing. This should be the #2 priority. But, along the lines of increasing the speed of Light, we need to find out exactly how Light propagates, then we can work to improve it. This would have such a large impact. I love all the little particles, but this is generations ahead of its time. As Engineers would say, "No bang for Buck." Someday it will be very important, but today we simply do not have the technology to make this useful. One day we will Engineer the Universe with these particles, but today, let's find Dark Matter/Dark Energy, or Let's nail the Ether, or Even Inflation. I love the little particles, but..

# GRAVITY

Okay, let's talk about space and the force of gravity. This is important because it is one of the building blocks of The Universe. Isaac Newton theorized that gravity was an attractive force between objects and the Earth. because of its size, Earth seemed to have very much. This theory was one of the centerpieces of Newton's Theories and was quite useful, but Einstein believed that something wasn't quite right. So he theorized that instead of Earth's gravity pulling objects to it, Space around the Earth was actually pushing objects towards Earth. If you believe Einstein, then the force of gravity may not exist. But they may be both right in a way. Einstein had the concept right, but did not take it to its logical conclusion. Never expanding the concept to replace Newton's Gravity with an all encompassing view of 3D space. If there is nothing in space, what is actually pushing down On the Earth and all the other objects in Space. He sort of defined what was going on around big objects, but never mentioned the rest of empty Space. Except to say that 3D Space existed. But how did it push? He proposed the fabric of Space and Time and large objects distorting that fabric. The problem is that nothing In Space is 2D and how do objects get pushed by a fabric that does not wrap all the way around and over a spherical object? Well, the answer is that Space is 3D and filled with a lattice like structure, much like the picture on the left.

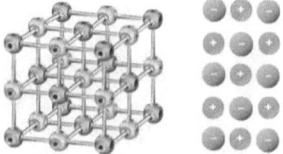

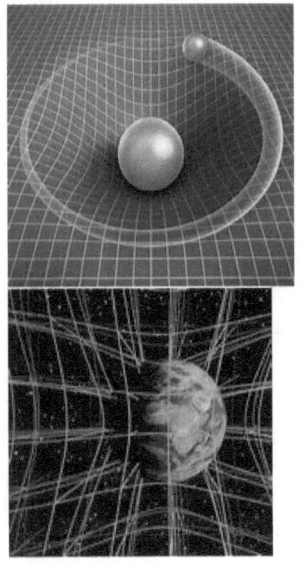

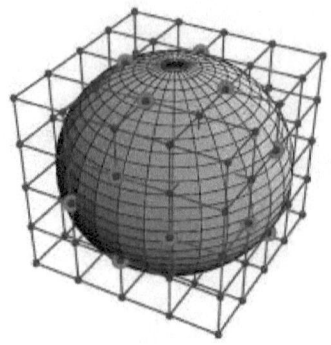

When an object is in Space it is surrounded by this matrix like the picture on the right, the matrix is distorted and attempts to return to normal. This force surrounds the object and pushes on it. Anything attempting to escape the object is pushed downward as well. Einstein called space 3D but failed to see that the fabric of Space is really a 3D Lattice or matrix of infinite size running thru Space. Before someone criticizes this concept, consider the Physicists easily accept the concept of a background radiation thru out the universe or even dust everywhere in the Universe, or my favorite, a curved Space. Infinite, but curved. Brilliant! Still no explanation of what is pushing down on me; Waves? , they have no mass, particles?, What pushes on the particles? Einstein was a master. Why not an Ether, that produces a Gravity like effect. This also helps to explain other things; Like how waves propagate thru space without a medium and how about missing energy and matter? Think it takes any mass or energy to build a 3D matrix in the whole Universe? Half of an Engineer's book and we are already on the way to unifying the universe! Just a quick note on the propagation of waves, such as light. We all know that sound waves need air to propagate and cannot be heard in a vacuum. Waves on the water need water to propagate, no water no waves. Shock waves can travel thru several materials, but need a medium. Warp drives as proposed, would create waves in Space to ride on. So why can physicists accept all these examples, but not a 3D Ether to carry light, or radio waves, or any kind of waves, AND support Einstein's gravity from all over the Universe. OK, enough common sense, back to Gravity. When an object is in this matrix it distorts the lattice and forces it to make room or wrap around the object. The lattice attempts to return to its orderly Shape but this stretching or distortion exerts a force 360 degrees around the object, causing a gravity like force that pushes objects toward the surface or accelerates them toward the surface. And, the bigger and massive the object, the more the matrix is distorted. So the gravity like effect is extended further out from the object. At that point far from the object, the lattice is able to return to its orderly shape. Einstein inadvertently proved this concept himself by showing that every point on the sphere of Earth had Space pushing objects down. The only way this works is if Space is 3D and surrounds Earth. But Space just doesn't surround Earth, it surrounds everything! Gravity works on the moon, the planets, the Sun, the whole Universe. And this leads to the conclusion that Space is one giant matrix that everything is interacting with. 3D Space, How could well educated Physicists actually believe that Space can be folded in two and connected with a worm hole when Space is 3D. Well, since Engineers don't like to use words like can't and never, maybe the matrix of 3D Space could be folded, but it ain't looking good. There are hundreds of images on the internet depicting SpaceTime, Worm Holes, Folded Space, Distortion of Space. Show me one that correctly shows 3D Space. That is exactly what Einstein called it, 3D AND how can gravity distort it unless it's made of matter.

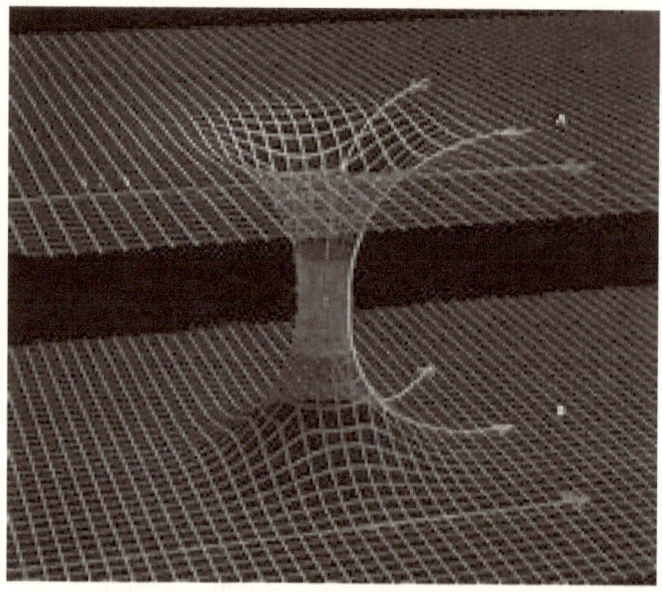

# Wormhole

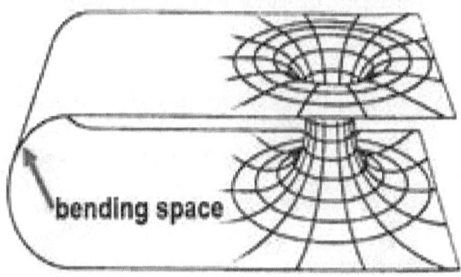

bending space

2D views of

Space and bending with 2D Wormholes.

Let't complete the Gravity thought. If an object is put into this Ether, it simply pushes the matrix out of the way. The matrix structure stays intact, but is compressed in the area around the object. The distortion is the greatest around the object, but the matrix is still distorted even at a distance from the object. And as you move further away, the distortion becomes less and less until none. This extended distortion is what causes other objects to orbit other objects, alter their course or simply fall into the larger object. So there is not Newton's gravity or Einstein's gravity or any gravity, there is only a 3D matrix running through the Universe causing a Gravity like effect. That, when distorted by an object, attempts to restore the original matrix structure, which causes the Matrix to exert a pushing action on the distorting object. No such thing as Gravity, but only an Ether or matrix that causes the effect. All that energy required to propagate a 3D matrix in Space; wonder were all that missing

Dark Energy is Physicists? Maybe right in front of your face! So to help Physicists to go from 2D Space to 3D Space, let me give you a picture of exactly how Gravity works in the 3D Ether and helps to push things down. First you must understand that the current model of the Universe believes that all mass objects carry a positive charge and Gravity a negative charge, the result is a zero charge Universe. So, Let's use that to our advantage for an Ether form of Gravity.

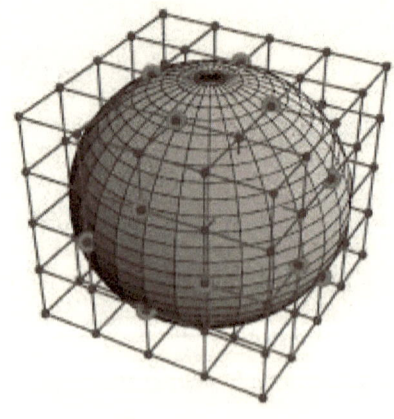

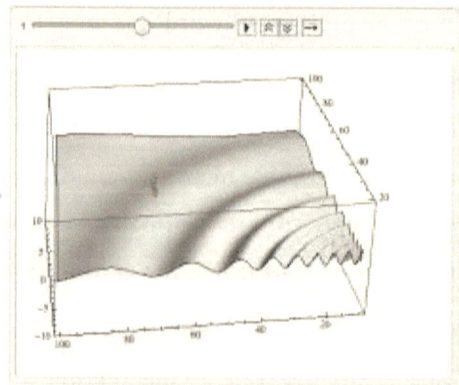

The picture on the top is a planet surrounded by the Ether, remember the Ether goes on to fill the entire Universe and it is negatively charged and produces gravity like effects. The picture on the bottom is a partial slice of the distortion caused by a large object in the Ether; more severe close to the planet, lower right, moving to less severe as you get further away from the planet, until there is no distortion (upper left of pic). So, this is where it gets exciting. If mass objects are considered positive and the Ether is negative; AND the area near the planet is more distorted with a higher negative charge, ANY Positive mass will be attracted to the distorted negative field, and further, since the distortion is most near the planet or more negative. The objects will be attracted all the way down to the planet, mimicking Gravity (an easy computer simulation). And I might add, that Gravity is a weak force, so don't expect positive cars to be lifted off the planet by the negative force. And remember the Ether has been stretched to fit around the planet and this stretching may act like a rubber band to keep objects on the surface and below. Now that we have established that some type of 3D matrix exists in space and works well with Einstein's Gravity theory, let's expand a little. NOTE: Physicists sometimes get hung up on words. So I would call 3D Space The Ether, but word definitions are not what I am trying to create. Call 3D space a matrix, a Lattice, the 3D Fabric of Space, a field, or 3D bubble wrap, Whatever! As long as we can all agree that there appears to be something out there. Now I have already alluded to the fact that folding an infinite 3D matrix would be extremely unlikely, so as well a localized folding would also be highly unlikely. In the form it was presented, An Einstein- Rosen Bridge is not possible and therefore incorrect. The Ether will not lend itself to being folded and a 2D fabric of Space just does not exist. I want to be very clear that although folding 3D Space does not appear possible, a worm hole may still be possible. But this means that you would simply be connecting point A In Space to point B in Space without the folding, which is no shortcut at all. What's the point! I am shocked that Physicists bought into a 2d Space Time, But it does go well with all the 2D planets and stars that they must believe exist.

 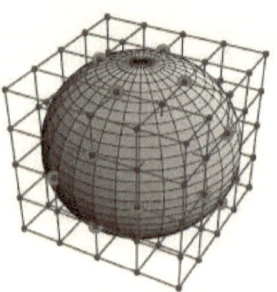

2D Space vs 3D Space.

# DARK ENERGY/DARK MATTER

This is my favorite subject because I have a chance to help find half the Universe.  How could anyone lose half of a Universe?  Wouldn't the first thing that would come to mind is that someone miscalculated The Universe is so large and there are so many zeroes after the numbers.  Also, as Physicists like to say, we are all made of star dust.  Tons and tons of it everywhere in the Universe.  And how about particles? We are up to 25 or so.  Again, all over the Universe. Maybe some are running around free.  Dark energy? The last chapter we establish that the Universe was full of a 3D matrix or Ether.  Could there be any energy in an Ether that "pushes down" on all objects in the Universe as Einstein theorized?     Ok, enough clues.  What Engineers like to do when not all data is available is to find the best data we can and project it into similar situations.  We have a very good estimate of the mass of our solar system, Sun, Planets and moons, asteroid belt, Keiper belt, dust and all.  Our Sun is considered a medium sized Star.  So if a star is twice as massive as our sun, give it double the mass of our solar system.  Continuing until the whole Universe is estimated

Review and adjust as needed.  Keep in mind that we know a little about stars but almost nothing about solar systems.  So our solar system model is the best place to start; we have a sun, planets, asteroid belt, moons, Gas Giants and solid planets, Kieper belt, dust. The mechanics of our solar system seem pretty normal. Of course other stars will be bigger or smaller, solar systems bigger or smaller, etc., but they should have the same mechanics. AS A FOOTNOTE: I feel as though I have to explain what I mean by mechanics Mechanics is the process of making something or how it moves, controlling it, even destroying it. Our Solar System seems to have things normal or average or by the rules. There is an average Sun, there are Solid planets, Gas planets, moons, dust, Keiper belt, Astoid belt, Gravity, pretty normal stuff.

## Main Sequence Stars

| Spectral Type: | O | B | A | F | G | K | M |
|---|---|---|---|---|---|---|---|
| Temperature: | 40 000K | 20 000K | 8500K | 6500K | 5700K | 4500K | 3200K |
| Radius (Sun=1): | 10 | 5 | 1.7 | 1.3 | 1.0 | 0.8 | 0.3 |
| Mass (Sun=1): | 50 | 10 | 2.0 | 1.5 | 1.0 | 0.7 | 0.2 |
| Luminosity (Sun=1): | 100 000 | 1000 | 20 | 4 | 1.0 | 0.2 | 0.01 |
| Lifetime (million yrs): | 10 | 100 | 1000 | 3000 | 10 000 | 50 000 | 200 000 |
| Abundance: | 0.00001% | 0.1% | 0.7% | 2% | 3.5% | 8% | 80% |

### Giant Stars
Low mass stars near
the end of their lives.

| | |
|---|---|
| Spectral Type: | Mainly G, K or M |
| Temperature: | 3000 to 10 000K |
| Radius (Sun=1): | 10 to 50 |
| Mass (Sun=1): | 1 to 5 |
| Luminosity (Sun=1): | 50 to 1000 |
| Lifetime (million yrs): | 1000 |
| Abundance: | 0.4% |

### White Dwarfs
Dying remnant of an
imploded star.

| | |
|---|---|
| Spectral Type: | D |
| Temperature: | Under 80 000K |
| Radius (Sun=1): | Under 0.01 |
| Mass (Sun=1): | Under 1.4 |
| Luminosity (Sun=1): | Under 0.01 |
| Lifetime (million yrs): | – |
| Abundance: | 5% |

### Supergiant Stars
High mass stars near
the end of their lives.

| | |
|---|---|
| Spectral Type: | O, B, A, F, G, K or M |
| Temperature: | 4000 to 40 000K |
| Radius (Sun=1): | 30 to 500 |
| Mass (Sun=1): | 10 to 70 |
| Luminosity (Sun=1): | 30 000 to 1 000 000 |
| Lifetime (million yrs): | 10 |
| Abundance: | 0.0001% |

r powell

This is he chart I used for these calculations. So, continuing If a star is half as massive as our Sun, give it half the mass of our Solar system and continue for all the lighter and heavier Stars. So, this is method yields:     ASSUMED – 100 billion galaxies, 10 trillion     billion stars     1 sun = 1.99E+30 kg,     1 solar system = 2.67E+27 kg,     10 trillion billion stars = 1E+21.

Mass Speadsheet part 1.

| UNIVERSE | | | | | | |
|---|---|---|---|---|---|---|
| +dark matter - | 3x10^55 | Size | | | Ave | |
| # of stars/class | % of suns/clas | (suns) | Class | | sol sys+sun mas: | mass x # of sol sys |
| 1E+14 | 0.00001% | 50 | O | | 9.96E+31 | 9.96E+45 |
| 1E+18 | 0.1% | 10 | B | | 1.99E+31 | 1.99E+49 |
| 7E+18 | 0.7% | 2 | A | | 3.98E+30 | 2.79E+49 |
| 2E+19 | 2% | 1.5 | F | | 2.99E+30 | 5.98E+49 |
| 3.5E+19 | 3.5% | 1 | G | | 1.99E+30 | 6.97E+49 |
| 8E+19 | 8% | 0.7 | K | | 1.39E+30 | 1.12E+50 |
| 8E+20 | 80% | 0.2 | M | | 3.98E+29 | 3.19E+50 |
| 1E+19 | 0.3% | 0.075 | L/T | | brown dwarfs | 7.55E+21 |
| | | | | | | |
| 4E+18 | 0.4% | 2.5 | Giant Stars | 4.98E+30 | | 1.99E+49 |
| 5E+19 | 5.0% | 1.3 | Wh Dwarfs | 2.59E+30 | | 1.29E+50 |
| 1E+16 | 0.0001% | 35 | Super Giant | 6.97E+31 | | 6.97E+47 |
| | 100.0% | | | | | |
| stellar black hol | assumed | 10 | | | left out | 1.39E+43 |
| galaxy black hol | assumed | 2E+09 | | | left out | 3.98E+53 |
| | | | | | | 7.58E+50 |
| | | | | | | MASS OF UNIVERSE |

Explaining these calculations: I simply took the number of stars in the Universe and divided them into the star sizes(column 1) using percentages from the chart.(column 2) Based on the ave. size(column 3) of a group of stars, I multiplied our sun's mass and Solar System mass(1.99E+30) by the scale factor of that group of stars(col 3) x the actual number of stars in that group(col 1), and that calculated the total mass of all the stars in that group(col 6). I now had the sun and solar system mass for that group or classification of stars. I then continued that process for every group of Stars. This gave me the basic mass of the Universe. That left a few extra items to be calculated, like Galaxy dust mass, or dust that has fallen on the solar systems,

and Black Holes.  Not included was Inter-galaxy and intra
Galaxy Space Junk; Dust, Asteroids, Rogue plants, Etc. Black
Holes I calculated from any est. I could find, but eventually left
it out as too speculative.

| | EARTH | Dust/Univ. mass | 98% empty space |
|---|---|---|---|
| | Dust/Yr Earth | % of Earth's Mass | ((Area(SS) x .98/Earth a |
| | 3311225.25Kg | 5.543098383E-19 | Earth Dust=SS Dust) x st |
| mass x # of sol sys | Dust/year | Star (SS)Mass x % | number in Universe SS |
| 9.96E+45 | | 5.52E+27 | left out |
| 1.99E+49 | | 1.10E+31 | left out |
| 2.79E+49 | | 1.55E+31 | left out |
| 5.98E+49 | | 3.31E+31 | left out |
| 6.97E+49 | | 3.86E+31 | left out |
| 1.12E+50 | | 6.18E+31 | left out |
| 3.19E+50 | | 1.77E+32 | left out |
| 7.55E+21 | | 4.19E+03 | left out |
| | | | left out |
| 1.99E+49 | | 1.10E+31 | left out |
| 1.29E+50 | | 7.18E+31 | left out |
| 6.97E+47 | | 3.86E+29 | left out |
| | | | 0.00E+00 |
| 1.39E+43 | | 7.72E+24 | **left out** |
| 3.98E+53 | | 2.21E+35 | |
| **7.58E+50** | | 2.21E+35 | |
| **MASS OF UNIVERSE** | | | |

Mass

Speadsheet part 2 Dust.

Dust, again, I picked the best data I could find.  This was an est.
of dust that fell on Earth every day, 40-60 tons per day.  I settled
on 10 tons, because I wanted only stardust.  I then multiplied that
by 365 days.  I knew that dust would hang in Space a long time,

so I gave it a year to settle. I truly believe this number is very light because dust has been produced since the big bang or 13.8 billion years ago. Anyways, I then took Earth's dust as a percentage of Earth's mass and applied it to our sun and solar system. Using the scale factors, I then applied the dust to all stars and solar systems, because dust would not just fall on Earth, it would fall on the whole solar system and sun, all across the Universe. This is where I struggled because I knew that matter was only about 2-4% of the solar system. I calculated dust mass on solar system mass, but dust would fall evenly even where there was no mass in Solar Systems. I also calculated the dust in the massless areas outside of Solar Systems, which was a much bigger number because there is only 2-4% mass. I ended up not including this number either for two reasons, it didn't seem like the right thing to do AND the open space Junk and the Black Hole masses had very little effect on the total mass of the Universe.     There is a number, but it would be WAG.  Doing the mass this way, yields a total mass number that is lighter than what Physicists calculated, but this is a spreadsheet that is based on the best mass numbers in the Universe, our Sun and Solar System. As long as it fits with an Ether concept to hit theoretical Universe mass, then it works.

This was all put on a spreadsheet so I could manipulate the numbers to fine tune the mass of the Universe.     After playing with every knob, masses, numbers of stars, number of Galaxies, sizes of stars, distribution of stars, all within reason and adding in Black Holes and heavy dust numbers, I could not come close to the theoretical $3\times10^{55}$ number.    That is, until I added in a 3D Ether calculation based on a matrix of electrons and protons in an 8 electron cube with a single Hydrogen atom in the center. Taking 1 cubic meter and dividing it into 64 smaller cubes each with a Hydrogen atom in the center gives the magic mass number, because of electron sharing with stacked cubes. 3.836E+55.    Dividing a 1 meter cube into 64 smaller cubes may seem like a lot, but actually each cube is approx. 80% of 1 foot by 1 foot by 1 foot. So I would think for bonding purposes it would be better to have smaller cubes, but the mass would go up. Maybe eliminating the proton from the center of the cube would allow for more electrons or a cloud, but would be less stable. And this would also prevent the use of Hydrogen as the building block. So, is this perfect? No, I chose this to meet a mass target. Could the cubes be spheres or other shapes, yes, absolutely. What the Ether actually looks like is TBD, but the concept of an Ether is solid. So, to be honest, I am not in love with this solution, my common sense alarm is going off. The particles should be smaller I feel. But the criteria is; must have mass to eliminate Dark Matter, must be negative in charge to use as Gravity, must have sufficient charge to use as Dark Energy and must cause the negative charge to increase towards the outside to explain Inflation. Easy Right? This does meet all the criteria and I will explain how this does it all: This is a Hydrogen atom on the inside, bonded to eight corner electrons, forming a cube 250mm and a charge of -8. The cubes will be stacked, causing electron sharing at all corners and the charge will drop to -1. To fill The Universe, another row of cubes will be put over this cube, this will make 3x3x3 cubes total and a -27 charge. The next row will take 5x5x5 cubes or 125 cubes with a -125 charge. As you can see, this increasing negative charge will fuel Inflation and pull matter to the outside of the Universe. Coninue this process and fill the Universe. This will solve Dark Matter/Energy and Inflation.

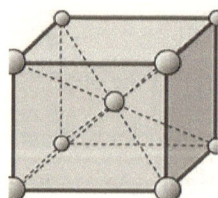 One comment about the lattices in nature. There are many known ways to arrange electrons and protons to form a lattice. I was trying to be as charge balanced as I could AND because protons are so heavy as compared to electrons, a less proton lattice was selected. With more protons like Carbon, Oxygen, Nitrogen, I could have easily exceeded the mass of the Universe by a factor of 100. Please see the Chapter on Inflation for a more detailed explanation on the Mechanics of the Ether.

Protonless bonding.

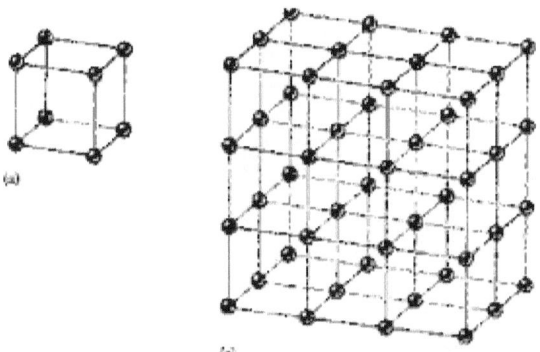

Tetrahedral bonding configuration

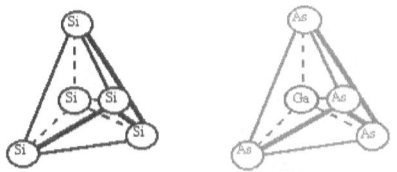

Face Centered Cubic and Diamond
Crystal Structure

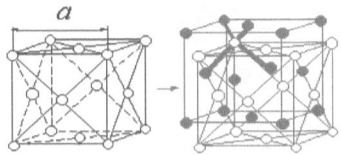

common types of bonds in nature:

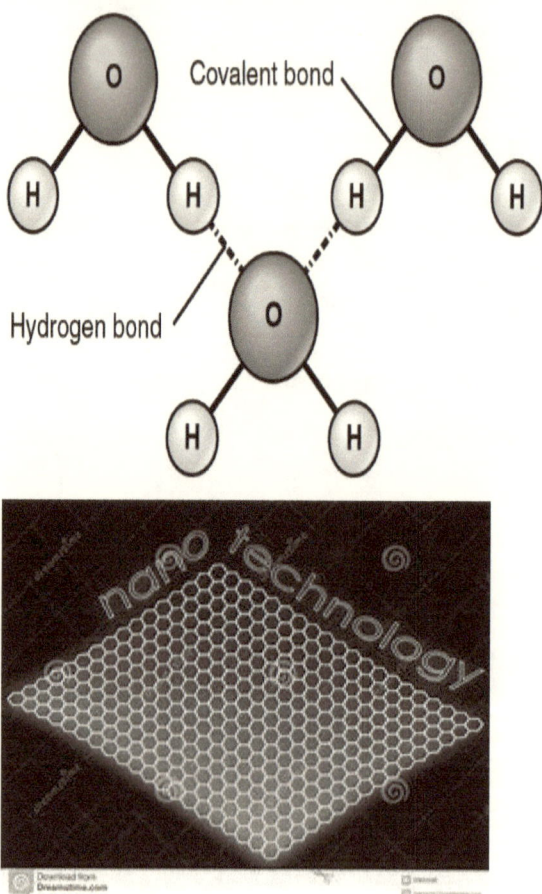

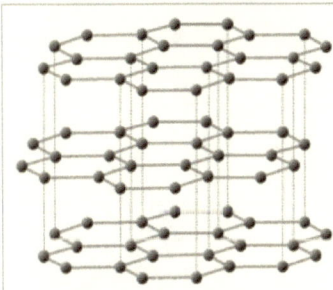

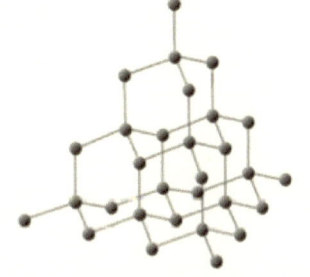

Graphite (solid lines are strong covalent bonds, dotted lines are weak inter-layer bonds)

Diamond (all bonds are strong covalent bonds)

Examples of bonding structures in Nature.

# INFLATION

I have talked about the Ether supporting the propagation of light and the WAG of how it imposes the Speed (Propagation) of light speed limit. I also talked about an Ether supporting a pushing Gravity. Objects in Space simply stretch it around the object (deforming), causing a gravity like effect. Now, for the "Hat Trick", I BELIEVE THAT THE ETHER IS CAUSING INFLATION, a WAG! Here is the mechanics of how that works: We are to fill the Universe with a cubical lattice Starting with 1 cube with 1 Hydrogen atom in the center with 8 electrons at the corners (shared with adjoining cubes) would produce a negative charge per cube and a slightly negative Ether. I say stacking cubes to help visualize how the negative charge grows as we stack more and more cubes around a center point. Actually, because of electron sharing, the cubes become a seamless matrix of electrons, like the picture on the right. To explain further, there is 1 cube to start with 8 electrons at the corners and a Hydrogen atom in the center, -8 charge, but as layering starts, electron sharing starts and the cube charge drops to -1. To make another layer of cubes around this 1 cube, it will take 27 cubes to achieve the second layer, and a -27 charge, already we can see the minus charge growing, -8 to -27.

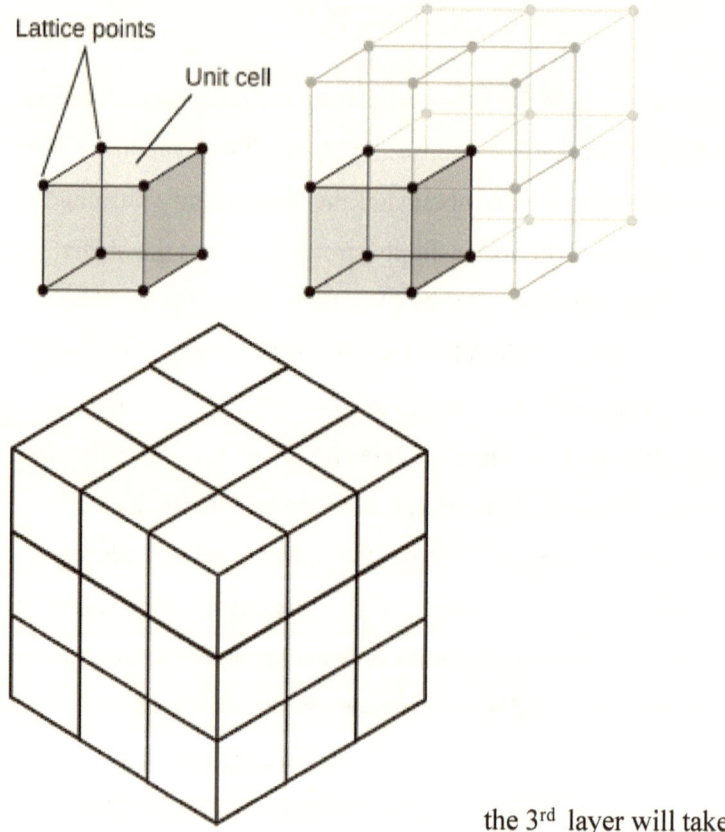

the 3$^{rd}$ layer will take 125 cubes to cover the second and so on. . .with each layer is

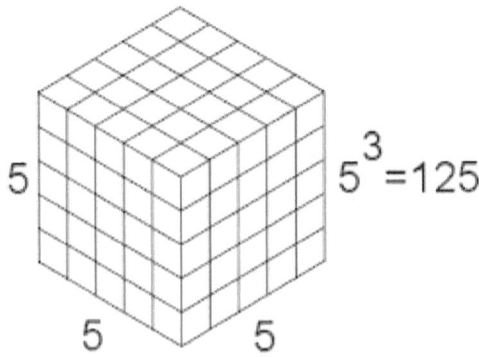

$$5^3 = 125$$

more

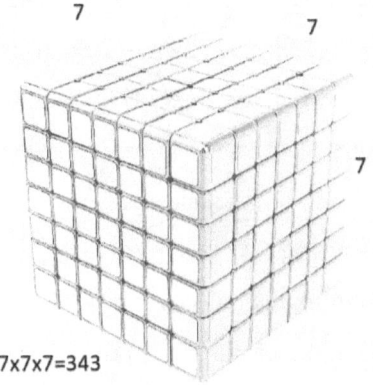

7x7x7=343

negative than the previous, layer 1 is -8, 2 is -27, 3 is -125, 4 is -343. This negative number also equals the number of cubes it takes to complete the next layer, because each cube has minus 1 charge. There are 8 electrons, 1 at each corner, BUT each electron is shared with 8 other cubes as they stack. So, -8 charge of 8 electrons gets divided by the number of shared cubes, which is 8 per electron. Each electron charge gets divided by 8, 1/8=-.125 charge to give to its cube. 8 electrons at the corners, each contributing -.125= a total charge for the cube of -1, as the hydrogen atom at the center is neutral. But wait you say, how does 8 electrons bond with a neutral atom? But the electrons charge is split with adjoining cubes and the split 8 electons act like the charge of just 1 electron. Again you say, but the Hydrogen atom is neutral, there is still to way to bond to a -1 charge. Ah, but there is young Grasshopper. Every Nucleus of atoms has orbitals where electrons circle the nucleus. These are organized into levels beginning with the closest to the nucleus. Because it is so close to the nucleus, it can only accept 2 electrons, any more electrons will have to go to the next orbital, which allows 8. Hydrogen is neutral, with 1 proton and 1 electron BUT has an opening in its first orbital for another electron. But, but the nucleus is neutral how can it bond with another electron? Ok, you must keep this a secret, because Physicists like to think they discover everything. The electron in the atom is orbiting the proton. The proton is as big as a house compared to the electron. As the electron orbits the big proton, the backside of the proton is exposed, think of it as the Dark side of the Moon. This gives a second electron an opportunity to slip in there and steal some of the proton's charge. This makes the bonds weaker and therefore les stable, but it works and in fact, is pretty common. Electrons are stripped away and gained allthe time. The Physicists even have a word for it, we have in affect just created an Anion. This can work, but don't tell the Physicists, they are busy looking for Dark Matter.

Hey, maybe we can lend them some. Enough fun, Continuing on.. .As the volume of the Universe fills with layer after layer of cubes, the outer layers become more and more negative. This is what attracts the positive matter towards the edges of the Universe at an increasing rate. (Again, I smell a computer simulation). The Universe cannot be zero energy, because nothing would move in or out. Now this is highly speculative, but an easy computer simulation. Oh, by the way, there is no Dark Anything just a 3D Ether. One of the reasons that the Ether Theory was abandoned was the belief that an Ether would allow speed to be additive. A light turned on from a moving plane should travel the speed of light plus the speed of the airplane. But light only traveled at the speed of light. Some Physicists believed that proved there is no Ether because an Ether would allow speed to be additive, but in fact, the complete oppose is true. The Ether has always been the limiting factor to electromechanical waves and will not support faster than light speed propagation, no matter how fast they could or should be going. The reason I bring this up here in the Inflation Chapter, is that the same reasoning is used for the Ether and Inflation only reversed. People think that an Ether should bring stability to the Universe, after all it limits the speed of Light doesn't it? Again wrong logic. The Ether only limits electromagnetic waves to the speed of light. Mass objects (galaxies, planets, my old socks) can go whatever damn speed they like So the Inflation of the

Universe (mass) is entirely possible with an Ether. And as I have shown, NOT ONLY IS IT POSSIBLE, BUT NEGATIVELY CHARGED ETHER INCREASINGLY NEGATIVE TOWARD THE OUTSIDE IS ACTUALLY CAUSING THE INFLATION! And, not only that, THERE IS NO SPEED LIMIT ON HOW FAST THE UNIVERSE CAN EXPAND!! Gravity, Propagation of the speed of light, Dark matter, Dark Energy, Inflation, Time manipulation, not too shabby for a first book. Physics is easy. Especially if I don't have to prove (Physical testing) anything. Ok, you talked me in to it, 1 more shot At the Physicists, oh, I mean let's do some common sense. If the Universe is expanding faster than the speed of light, that must mean that either something is moving faster than light OR multiple objects are moving and their speed is additive to match the inflation. First, Einstein won't allow mass objects to go faster than the speed of Light, that infinite mass infinite Energy thing, so that's out (Wrong, but next book). Then it must be the additive speed thing, as The Theory of Relativity says that two objects moving away from one another is the same as one being stationary and the other appearing to be moving faster than light, which Einstein says is not possible. It looks like Einstein has you boxed into a corner Astronomers. Either Einstein is wrong (nothing can move faster than light) or Relativity( Einstein ) is wrong, or The Astronomers are wrong

and inflation is  not as fast as they say.  So many flaws. I am
pulling  for the Astronomers, I want to go faster than Light.

# THE TWIN PARADOX

## (Or This is Quite Possibly the Dumbest Thing I Have Heard in Physics)

Even though these are related, I will take one at a time. I really only wanted to write this book to show where Dark Energy and Dark Matter were. I like Einstein, I like Newton, I even like most Physicists, but sometimes I just can't figure out what they are thinking. OK, the twin Paradox. One twin stays on Earth and becomes an Engineer. The other twin becomes a Physicist and goes into space and flies near the speed of light for a year then returns home. Upon landing, the space traveling twin meets with his brother and declares, "Brother I have been traveling near the speed of light for a year. Everyone knows time goes slower near the speed of light. I am now about 1 hour younger than you!" The Engineer looks at his brother and says "I love you brother, but sometimes you are an idiot! Who in their right mind would want to fly at the speed of light for a year, not being able to see anything passing by. Now you claim that you are younger. What were you smoking in that spaceship?" "Look!" said the space brother pulling out a clock from his bag, "We synchronized our clocks before the flight. Now, my clock is 1 hour behind yours!" "Really?" said the Engineer, "maybe it's broken." "No way!" said spaceman, "it has been running perfectly ever since I landed, Remember these 2clocks were the best money could buy, Certified for accuracy by you Engineers, I am younger!" "Ok," said the Engineer, "I will explain this to you slowly." "There is only one Clock in the Universe for people living on Earth, and it is called the Earth. Time is derived from its rotation and its rotation around the Sun. Your clock might disagree, but the Earth is never wrong. Your clock may have been affected by the speed of light or lack of Gravity, radiation, heat, vibration, pressure, but for some reason it is wrong. Even if you run this test the same way over and over, the clock is wrong. The Earth kept the same rotations all while you were gone. Time is not kept by a clock on your spaceship, it is kept by Earth. And it says you are just as old as I am. There is no paradox. Speed has nothing to do with time. 0 miles an hour or 2 million miles an hour, an hour is an hour. Brother, if the clock on your ship stopped would you go into suspended animation until the clock was fixed? No, of course not brother and you know why? Because you and I and 7 Billion other people live by the Earth's Clock, it is our Reference Frame for Time, don't let someone change your Reference Frame (ship's Clock) and make you think you are Younger. You are, but only in their Reference Frame. Maybe they were trying to sell a Book or a Video or maybe even one of those Nobel Prize things. So, when a master magician like Einstein says "watch me pull a rabbit out of my hat! You have to see the Big Picture. Know you're Frame of Reference. Einstein switched Frames of Reference to one where

Clocks can slow down" "Not clocks, but Time itself" interrupted the brother. "Don't you see?" said the Engineer "That is his Reference Frame and he Can do whatever he wants, I love you, but you're an idiot!" "No, No, they did Real Tests! You can pull a rabbit out of a hat! Time Dilation does exist." Said space brother. "Okay, you never did know when to quit. Three things before I quit. 1. This is why Mom always liked me best! 2. This genius also came up with Gravity Dilation, which is kind of the opposite of Time Dilation. Clocks run faster the higher up you go. I can tell by the way you are thinking, You must have been really high, so how fast were you actually flying? Yes I know you tell everyone you were traveling near the speed of light, but really, how fast were you going? What? 30,000 MPH? So Time Dilation had little effect on you, but the bad news for you is that Gravity Dilation had more effect at that speed, you are really older, not young, than me!" "Wait!" said spaceman, "you said there were 3 things. What is the third?" "Well, I knew you were not flying Near the speed of light, because your friend, the genius, also said that mass objects become more Massive as they approach the speed of light, and this takes more and more energy. Sorry, you can't get there from here. Therefore as proposed, Time Dilation and Length Contraction can't exist. I have seen Theories be wrong, but I have never seen someone discredit their own Theory. But it's okay, Brother, Watch me pull a rabbit out of my hat!" What are Physicists thinking? To be valid on Earth, you can only use the Earth's master Clock, no other. If you use a Clock and it does not agree with the master, it is wrong OR you are switching Reference Frames. If Time actually slowed down, Earth's Clock would slow down. Wake up! Earth's Clock, the agreed Master, shows Zero evidence of Time Dilation. You have to decide if you are going live by Earth's time or some other Clock somewhere.

# RELATIVITY

Well, here we go again. Same stuff, different subject. Einstein
was a master at manipulating Physical properties to support his
Theories. As in the Twin Paradox previous, Einstein tries to
manipulate time and motion to support Relativity. As pointed
out, time is a constant. Fixed by the rotation of the Earth, so
when you hear someone talk about slowing time, your little BS
Detector should be sounding an alarm. A second In space will
be a second on Earth, an hour is an hour. When someone
proposes adjusting time, it really means they don't have a clue to
the real answer. I don't want to go thru and correct Relativity
(Ok, next book) and I feel like I have already beat Time Dilation
to death. So for now, just let me say, in the Earth's Reference
Frame, there is a master Clock that does not speed up, slow
down or stop. Changing to a different Reference Frame to
manipulate Time means nothing to us using Earth Time. End
Of Story.

# EPILOQUE

As I was writing this book and doing research, I was truly impressed by the amount of work that had been done and is being done by Physicists. Of course, as an Engineer, I would like to see more data and more Physical testing with more precision. Speculation is fun and interesting, but mostly it clouds the real answer. Anyone can say anything. Require a certain level of Physical proof, and all the alternate Theories go away. Instead of rehashing or challenging old Theories, we could all move forward. Which brings up the next flaw, why is there not an Academy of Physics that is a clearing house for all approved Theories and data. Just finding information took hours. I couldn't tell what was an accepted Theory or just speculation (String Theory). I thought a Theory had to have some kind of proof? Who approves the Physics use of the term "Theory"? Or is this all just made up? Anyways, I learned a lot about Physics and can speak the language a tiny bit better. And the search for the mysteries was kind fun (what a Liar!). It was touch and go at some points, but as promised, I will write another book on the Universe from an Engineering Standpoint. I kid about String Theory, but I see some potential. I promised I will show a way to go faster than light (What??), and Warp drives need to be looked at. Anti-matter anti-particles need a feasibility look. What is a Joule and why does it exist? Should be replaced by something "cooler". Oh, I am avoiding this

Elephant in the Room, but particle/quantum Physics seems like a runaway freight train, 30 particles to build my Universe? 5 tops! My BS detector is starting to vibrate. And let's correct that silly Double Slit Experiment to Reality. I know that cutting a couple of slots in a piece of cardboard and shining a flashlight at it is quite an accomplishment for Quantum Physicists, but REALLY, REALLY! Not prejudging I will keep an open mind, but if I had a choice I would have taken the money spent for the CERN Collider and built the biggest, baddest, Telescope in the Universe. Imagine the pictures and the data the Astronomers would produce. Then compare that to the pictures being produced at the CERN Collider. And why does every comment from particle Physicists end like, "We are just 1 more particle away from an answer!" (See Cern comments in publications). Neutrinos can go thru solid matter unimpeded? (right, and pigs fly!)

And why is there, all of a sudden, a massless particle in a Photon. What's the point? Don't you know that $E=mc^2$ thing? Put a zero in for mass(m) and E=0, you got nothing, you went from infinite mass at the speed of light, to nothing at anything!! Good move. And F=ma works well with zero mass. E=0, F=0, you guys are brilliant! I am just kidding, Physicists. Keep trying. Anyways, There is plenty of material for another book, maybe an Encyclopedia, I just have to rest first. So many flaws, so little time. Oh, and one last thing. First, I will leave Einstein alone, believe what you must. And just to end On a high note, I have a Certainty Principal. Oh, and the Universe appears to be flat and not Expanding quite equally, why? I have a Theory! To be continued. . . .

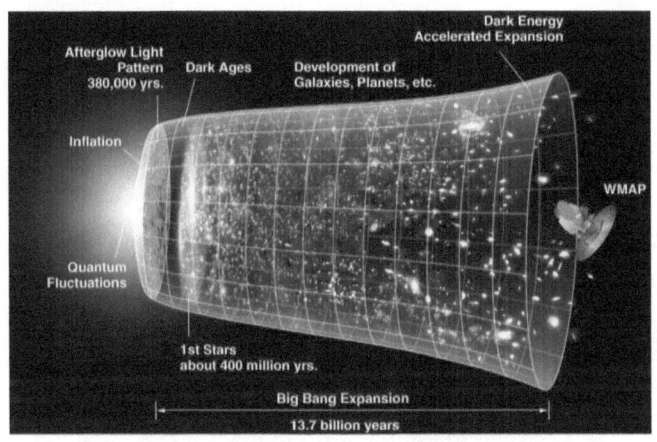

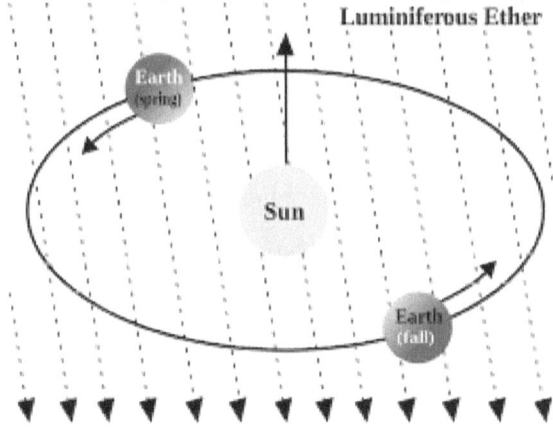

# APPENDICES

Had to include this from the internet, as sometimes Physics isn't real clear. **c=The Speed of Light** Although $c$ is now the universal symbol for the speed of light, the most common symbol in the nineteenth century was an upper-case $V$ which Maxwell had started using in 1865. That was the notation adopted by Einstein for his first few papers on relativity from 1905. The origins of the letter $c$ being used for the speed of light can be traced back to a paper of 1856 by Weber and Kohlrausch [2]. They defined and measured a quantity denoted by $c$ that they used in an electrodynamics force law equation. It became known as Weber's constant and was later shown to have a theoretical value equal to the speed of light times the square root of two. In 1894 Paul Drude modified the usage of Weber's constant so that the letter $c$ became the symbol for the speed of electrodynamic waves. In optics Drude continued to follow Maxwell in using an upper-case $V$ for the speed of light. Progressively the $c$ notation was used for the speed of light in all contexts as it was picked up by Max Planck, Hendrik Lorentz and other influential physicists. By 1907 when Einstein switched from $V$ to $c$ in his papers, it had become the standard symbol for the speed of light in vacuum for electrodynamics, optics, thermodynamics and relativity. Weber apparently meant $c$ to stand for "constant" in his force law, but there is evidence that physicists such as Lorentz and Einstein were accustomed to a common convention that $c$ could be used as a variable for velocity. This usage can be traced back to the classic Latin texts in which $c$ stood for "celeritas" meaning "speed". The uncommon English word "celerity" is still used when referring to the speed of wave propagation in fluids. The same Latin root is found in more familiar words such as acceleration and even celebrity, a word used when fame comes quickly. Although the $c$ symbol was adapted from Weber's constant, it was probably thought appropriate for it to represent the velocity of light later on because of this Latin interpretation. So history provides an ambiguous answer to the question "Why is $c$ the symbol for the speed of light?", and it is reasonable to think of $c$ as standing for either "constant" or "celeritas". But not Causality as some Physicists improperly refer to it. **The Michelson-Morley Experiment -** A Michelson interferometer uses the same principle as the original experiment. But it uses a laser for a light source. The Earth travels very quickly (more than 100,000 km per hour) around the Sun.[1] If aether exists, the Earth moving through it would cause a "wind" in the same way that there seems to be a wind outside a moving car. To a person in the car, the air outside the car would seem like a moving substance. In the same way, aether should seem like a moving substance to things on Earth. The interferometer was designed to measure the speed and direction of the "aether wind" by measuring the difference between the speed of light traveling in different directions. It measured this difference by shining a beam of light into a mirror that was only partially coated in silver. Part of the beam would be reflected one way, and the rest would go the other. Those two parts would then be reflected back to where they were split apart, and recombined. By looking at interference patterns in the recombined beam of light, any changes in speed because of aether wind could be seen. They found that there was in fact no substantial difference in the measurements. This was puzzling to the scientific community at the time, and led to the creation of various new theories to explain the result. The most important was Albert Einstein's special theory of relativity. BUT WHAT THEY REALLY DISCOVERED WAS THERE WAS NO ETHER WIND. The test results were never corrected.

ADDITIVE SPEED                               Young Physicist

Students answering                 Physics questions It's

because the luminiferous aether was, by definition, composed out of some particles or elementary building blocks with a well-defined location in space. Consequently, it picks a privileged reference frame, the rest frame of the aether. In this rest frame, the speed of light – vibrations of the aether – could be constant, $cc$. However, things moving relatively to this aether by the speed $vv$ should detect a different speed of the light relatively to them – the speed would go from $c-vc-v$ to $c+vc+v$, depending on the direction. However, this modification of the light speed,

the so-called aether wind, was shown to be non-existent by the
Morley-Michelson experiment which measured the speed to be
$c$c regardless of the source and the observer. This falsifies the
existence of the aether. This is still being taught even though we
now know there is no mass in light therefore cannot be affected
by wind  *WHY ARE ALL MASS VELOCITIES IN THE
UNIVERSE ADDITIVE BUT LECTROMECHANICAL WAVES
ARE NOT?? BECAUSE ELECTROMECHANICAL WAVES
ARE PROPOGATED ALONG THE ETHER AND MASS
OBJECTS ARE NOT. WHO SAID THE ETHER COULD
PROVIDE FOR FASTER THAN LIGHT PROPAGATION,
PLEASE NAME A NAME. THE ETHER IS WHAT MAY LIMIT
THE SPEED OF PROPOGATION. I CAN FIND NO
EVIDENCE THAT ANYONE IN PHYSICS EVER BROUGHT
UP THIS ISSUE AND CERTAINLY NEVER TESTED FOR IT.
THEY WERE SO INTENT ON PROVING THAT NOTHING
COULD GO FASTER THAN THE SPEED OF LIGHT THAT
THEY FORGOT TO ASK, WHY? WHY? WHY NOT!! ( LET'S
JUST KILL THAT DAM\* AETHER THEORY!) LONG LIVE
SPECIAL RELATIVITY!* The equivalent but even more robust
refutation of the aether came from the theory. A physicist named
Albert Einstein built a whole new theory of spacetime, the so-
called special theory of relativity (a picture of this physicist is
often being shown by the ordinary people as well), that also
assumes/guarantees that the speed of light is always constant and

there can't be any privileged reference frame. Relativity has been backed by the Morley-Michelson experiment as well as hundreds of much more specific experiments. *MORLEY-MICHELSON EXPERIMENT DOES NOT BACK RELATIVITY AND IS A VERY FLAWED EXPERIMENT. See Aether chapter.* One of the things it guarantees is that light (electromagnetic radiation) has to be made out of disturbances of the empty space, the vacuum itself, and not a localized material carrier. DISTURBANCES IN EMPTY SPACE? WHAT IS THERE TO DISBURB? (name of contributor withheld to protect their intelligence or lack thereof) Who is teaching these Physics students? So, in rebuttal, Here is what Einstein said about "a localized material", *According to the general theory of relativity, space without ether is unthinkable; for in such space there not only would be no propagation of light, but also no possibility of existence for standards of space and time. But this ether may not be thought of as endowed with the quality characteristic of matter, as consisting of parts ('particles') which may be tracked through time.*

*(Albert Einstein, 1928, Leiden Lecture)*

*WHAT?? AN ETHER*

### SCIENTIFIC METHOD AND FLAWS

Since the 17th century, the scientific method has been the gold standard for investigating the natural world. It is how scientists correctly arrive at new knowledge, and update their previous knowledge. It consists of systematic observation, measurement, experiment, and the formulation of questions or hypotheses. 1. Formulate Question/Hypothis 2. Define the Research Question 3. Review the Literature 4. Create a Hypothesis 2. Collect Data 2. Preparation: Make the Hypothesis Testable 3. Preparation: Design the Study 4. Conduct the Experiment or Observation 3. Test Hypothesis 2. Organize the Data 3. Analyze the Results 4. Check if the Results Support or disprove the Hypothesis 4. Conclusion 2. Look for Other Possible Explanations 3. Generalize to the Real World 4. Suggestions to Further Research ------------------------------------------------------------------ --------------------- I included this only this to readers how science is supposed to be done. I don't have A problem with Astronomers, or strangely enough Quantum/Particle Physics, they Started out as the wild bunch, but now seem to be on a quest for the truth, And other Physicists. The group that has gone astray, like Einstein, are the Theoretical Physicists. Relativity is a thought experiment. Special Relativity is a thought experiment. How does Thought experiments become part of mainstream anything. Look at the Scientific Method. Deep-field images from the Hubble Space Telescope suggest there are 10 times more galaxies in the universe than scientists previously thought, with about 2 trillion galaxies in total, according to a study published in October 2016 in the journal Science by Christopher Conselice, a professor of astrophysics at the University of Nottingham in the U.K., and his colleagues. [Video: Our Universe Has Trillions of Galaxies] About 100 million (or 10 to the eighth power) stars inhabit the average galaxy, according to one of the best estimates, Conselice wrote in an email to Live Science.

# GLOSSARY

**TERMS** used in this Book

**A WAG** is a Wild *ss Guess – no physical data, weak/no logic, no certified test results, disagreement among peers. Only observable proof or None. A AG is **an Average Guess** – some basis, a little data, average logic, a good chance of being right. A UHG is an **Unusual Guess** – something different or new. Maybe thought of before but not presented. Most of my guesses here.

**A FACT** – Good Logic, physical data, Certified test results, much agreement of peers, works well with other Theories, very rare. Because of working with such big or small numbers, A Fact is very difficult. But that is never an excuse to approve something with No data. It just means it will take more work and creativity. Pack a lunch.

**A FLAW** – a mistake or some kind or error, even in logic or judgement.

**SIDEWAYS THINKING** – Something that is not completely or well thought out and may contradict itself.

So, It is only fair that if a few Physicists won't recognize the contributions of Engineering, it is only fair for Engineering (Me) to point to the flaws we see in Physics: